James Croil

Steam navigation, and its relation to the commerce of Canada and the United States

James Croil

Steam navigation, and its relation to the commerce of Canada and the United States

ISBN/EAN: 9783743467132

Manufactured in Europe, USA, Canada, Australia, Japa

Cover: Foto ©berggeist007 / pixelio.de

Manufactured and distributed by brebook publishing software (www.brebook.com)

James Croil

Steam navigation, and its relation to the commerce of Canada and the United States

Steam Navigation

AND

ITS RELATION TO THE COMMERCE OF CANADA AND THE UNITED STATES.

BY

James Croil,

MONTREAL.

AUTHOR OF "DUNDAS: A SKETCH OF CANADIAN HISTORY."

With Illustrations and Portraits.

TORONTO:
WILLIAM BRIGGS.
MONTREAL: THE MONTREAL NEWS COMPANY, LIMITED
1898.

ENTERED according to Act of the Parliament of Canada, in the year one thousand eight hundred and ninety-eight, by WILLIAM BRIGGS, at the Department of Agriculture.

This Volume

is dedicated by permission to

His Excellency the Earl of Aberdeen,
K.T., G.C.M.G., etc.,
Governor=General of Canada

from 1893 to 1898,
a nobleman who will long be gratefully remembered
as the benefactor and friend
of all classes of the community, and
who, with his Consort,

The Countess of Aberdeen, LL.D.

will always be associated by the
Canadian people with a period in their history of
great national prosperity,
their joint efforts in furthering lofty ideals
having done much to
advance the highest interests of the Dominion.

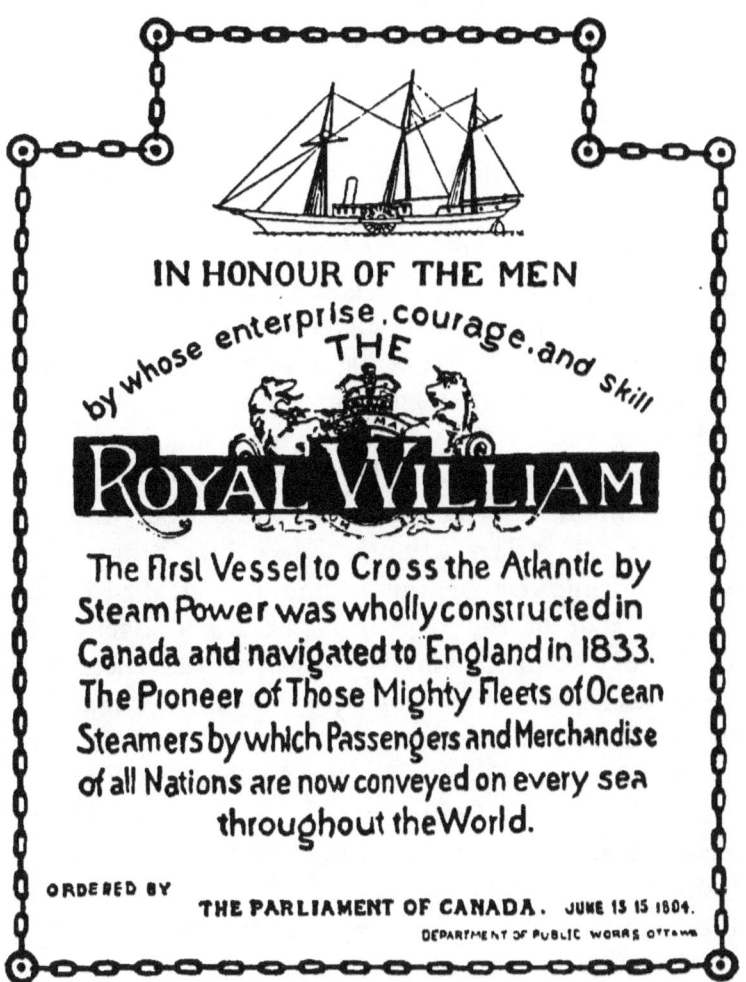

FACSIMILE OF THE MEMORIAL BRASS,

"ROYAL WILLIAM."

PREFACE.

WHEN the history of the nineteenth century comes to be written, not the least interesting chapter of it will be that which treats of the origin, the development, and the triumphs of Steam Navigation—that mighty combination of inventive genius and mechanical force that has bridged the oceans and brought the ends of the earth together.

During the past few years several important contributions to this class of literature have issued from the metropolitan press. Three of these deserve special mention: (1) "The Atlantic Ferry; its Ships, Men, and Working," by Arthur J. Maginnis, gold medallist and member of the Institution of Naval Architects, 1892; (2) "Our Ocean Railways, or the Rise, Progress, and Development of Ocean Steam Navigation," by A. Fraser-Macdonald, 1893; (3) "The History of North Atlantic Steam Navigation, with Some Account of Early Ships and Shipowners," by Henry Fry, ex-President of Dominion Board of Trade of Canada and Lloyd's Agent at Quebec, 1896. Each of these writers, in his own way, has treated the subject so thoroughly and satisfactorily, the author feels as though the wind had been taken out of his sails somewhat, and it is not without hesitation that he has yielded to the advice of friends in whose judgment he has implicit confidence, and ventured to follow in the wake of such accomplished writers.

If I am questioned as to *motif* I cannot better justify the rash deed than by endorsing the sentiment in Byron's apostrophe :

> "And I have loved thee, Ocean ! and my joy
> Of youthful sports was on thy breast to be
> Borne, like thy bubbles, onward ; from a boy
> I wantoned with thy breakers—they to me
> Were a delight."

These pages are of a much less pretentious character than the above-named books. They are but a compilation of materials more or less intimately connected with Steam Navigation, gathered from many sources, during many years, and now woven into homely narrative. They necessarily contain much in common with these other writings on this subject, but they are projected from a different standpoint and embrace a wider field, supplying information not easily obtained, respecting the far-reaching waterways of Canada, her magnificent ship canals, and the vast steam commerce of the Great Lakes.

So numerous are the sources of information drawn upon, it is impossible to make adequate acknowledgment of them all. The agents of Atlantic lines of steamships were particularly obliging in their replies to inquiries made of them. Without in any way making them responsible for the use made of their communications, upon these my remarks on that branch of the subject are chiefly based. Among other publications I have consulted the "Transactions of the Imperial Institute," London, and of the Literary and Historical Society of Quebec; Government reports emanating from Ottawa and Washington; also many pamphlets, magazine and newspaper articles bearing on the subject, not to speak of my capacious scrapbook and some well-thumbed note-books.

Additional authorities will be indicated as the narrative proceeds. Besides these, grateful acknowledgments for valuable assistance are due to Sir Sandford Fleming and Mr. George Johnson, F.S.S., of Ottawa; to Messrs. Douglas Battersby, R. W. Shepherd, and the late Captain Thomas Howard, of Montreal; to Mr. Archibald Campbell, of Quebec; Captain Clarke Hamilton, of Kingston; Mrs. Holden, of Port Dover, Ont., and Mr. T. M. Henderson, of Victoria, B.C.; to members of the Boards of Trade in Montreal, Minneapolis and Duluth; and to the following clergymen: Rev. Dr. Bruce, of St. John, N.B.; Rev. T. F. Fullerton, of Charlottetown, P.E.I.; Rev. James Bennett, of L'Orignal, Ont., and Rev. W. H. L. Howard, of Fort William, Ont.

The illustrations have nearly all been made for this work: the wood-cuts by Mr. J. H. Walker, and the half-tones by the Standard Photo-Engraving Company, Montreal.

J. C.

MONTREAL, *October*, 1898.

CONTENTS.

CHAPTER I.
THE DAWN OF STEAM NAVIGATION - - - - - - 17

CHAPTER II.
EARLY YEARS OF STEAM NAVIGATION - - - - - 50

CHAPTER III.
THE CUNARD STEAMSHIP COMPANY - - - - - 71

CHAPTER IV.
NORTH ATLANTIC STEAMSHIP COMPANIES - - - - 103

CHAPTER V.
STEAM TO INDIA AND THE EAST - - - - - - 142

CHAPTER VI.
STEAM IN THE BRITISH NAVY - - - - - - 166

CHAPTER VII.
THE ST. LAWRENCE ROUTE - - - - - - - 192

CHAPTER VIII.
STEAM ON THE GREAT LAKES - - - - - - 244

CHAPTER IX.
STEAM COMMERCE OF THE GREAT LAKES - - - - 268

CHAPTER X.
STEAM NAVIGATION IN ALL THE PROVINCES OF THE DOMINION
AND IN NEWFOUNDLAND - - - - - - - 307

ILLUSTRATIONS.

STEAM VESSELS.

	PAGE		PAGE
ALBERTA	285	NORMANNIA	131
ATLANTIC	105	NORTH-WEST	273
AUGUSTA VICTORIA	133	OCEANIC	117
BEAVER	335	OHIO STEAMER	45
BRITANNIA	72	PARIS	107
CALEDONIA	146	PARIS DINING-ROOM	109
CAMPANIA	78	PARIS *(Stern View)*	108
CANADA	226	PARISIAN	204
CHARLOTTE DUNDAS	32	PASSPORT	327
CLERMONT	42	PENNSYLVANIA	135
COLUMBA	38	PILGRIM	16
COMET	35	PRINCETON	253
CORONA	329	PRISCILLA	46
CRESCENT	191	QUEBEC	311
DUKE OF WELLINGTON	167	QUEEN CHARLOTTE	249
EMPIRE	255	QUETTA	150
EMPRESS OF JAPAN	162	RENOWN	172
GREAT BRITAIN	62	RHINE STEAMER	39
GREAT EASTERN	63	ROBERT GARRETT	49
HORNET	169	ROYAL WILLIAM	8
JEANIE DEANS	51	ST. LOUIS	111
JOHN S. COLBY	363	SAVANNAH	53
KAISER W. DER GROSSE	137	SCOTIA	77
LAKE ONTARIO	230	SIRIUS	59
MAJESTIC	119	SOVEREIGN	317
MANITOU	271	STANLEY	352
MILLER'S TWIN BOAT	31	TEUTONIC	174
MISSISSIPPI STEAMER	43	VANDALIA	251
NELSON	337	VICTORIA AND ALBERT	184
NEW YORK	47	WALK-IN-THE-WATER	250
NIAGARA	74	WILLIAM IV.	325

PORTRAITS.

	PAGE		PAGE
AIRD, CAPTAIN	215	McMASTER, CAPTAIN	197
ALLAN, SIR HUGH	208	McLENNAN, HUGH	296
ALLAN, ANDREW	296	MOUNTSTEPHEN, LORD	4
BURNS, SIR GEORGE	93	NAPIER, ROBERT	97
CAMPBELL, CAPTAIN	233	NAPIER, MRS.	97
CUNARD, SIR SAMUEL	93	OGILVIE, W. W.	296
DUTTON, CAPTAIN	218	RITCHIE, CAPTAIN	216
FLEMING, SIR SANDFORD	4	SHEPHERD, R. W.	322
GRAHAM, CAPTAIN	211	SMITH, CAPTAIN W. H.	194
HAMILTON, HON. JOHN	331	STRATHCONA, LORD	4
LINDALL, CAPTAIN	223	TORRANCE, JOHN	308
MACAULAY, CAPTAIN	227	WYLIE, CAPTAIN	212
MacIVER, DAVID	93		

MISCELLANEOUS.

	PAGE		PAGE
CANAL LOCK, CANADIAN	264	HORSE-BOAT	29
CANAL LOCK, U. STATES	278	MAP GULF PORTS, ETC.	241
CUNARD TRACK CHART	90	ROYAL WILLIAM—MODEL	55
GRAIN ELEVATOR	289	SHIP OF THE DESERT	143
GREAT REPUBLIC, SHIP	26	WIND-BOAT	70

"PILGRIM,"
Sister to *Priscilla* of the Fall River Line, 1890.

CHAPTER I.

THE DAWN OF STEAM NAVIGATION.

Ah! what pleasant visions haunt me
As I gaze upon the sea!
All the old romantic legends,
All my dreams come back to me.
—LONGFELLOW.

The up-to-date standard—Old-time sailing ships—The clipper packet-ship—Dawn of steam navigation—Denis Papin on the Fulda—Bell's *Comet* — Fulton's *Clermont* — American river steamers and ferry-boats.

TRAVEL increases in faster ratio than do facilities for inter-communication. The prophecy surely is being fulfilled in these latter days, "Many shall run to and fro, and knowledge shall be increased." It is estimated that at least 750,000 persons travel yearly between Europe and America; 99,223 cabin passengers and 252,350 steerage passengers landed at New York from Europe in 1896. The Cunard Line brought the largest number of cabin passengers, 17,999, from Liverpool, and the North German Lloyd Line the largest number of steerage, namely, 38,034, from Bremen.

Notwithstanding the wonderful development of railway and steamship systems, means of conveyance during the summer months often fall short of the demand. Passages by the more popular lines of steamships must be engaged months ahead; in many cases the ships are uncomfortably crowded. At such times sofas take the place of berths, and all the officers' rooms, from the coveted Captain's cabin to the second and third stewards' bunks, are called into requisition and held at a round premium. On Saturday, the 8th of May, 1897, no less than 1,500 saloon passengers left New York for Liverpool on the great ocean greyhounds. The travelling season is comparatively short, the competition is keen, and the enormous expense of building, furnishing and running up-to-date steamships renders it difficult to provide the requisite accommodation on a paying basis. The up-to-date steamship must be built of steel, to combine light weight with strength. It must have triple or quadruple expansion engines to economize fuel. It must be propelled by twin or triple screws, as well for the easier handling of the vessel as for safety in case of a break-down of machinery, and for attaining the highest possible speed. Our ideal steamship must be able to turn quite round in its own length, and to go through the water at an average speed of at least twenty knots an hour. To attain these results, ships of a very large class are called for—nothing short of from eight to ten thousand tons burthen will come up to the mark. There are many magnificent steamships in the North Atlantic trade and elsewhere but as yet few have in all respects reached the up-to-date standard, and even those that are such this year, a few years hence are certain to be regarded as quite behind the times. There is no valid reason to suppose that the process of development which has been going on during the last fifty years in this direction is to be arrested at the close of the century. The indications, so far as they can be interpreted, are all in the opposite direction. The paddle-wheel ocean steamer reached its zenith with the launch of the *Scotia* of the Cunard Line in 1862. She was the last of the race.

The wooden steamship, "copper-fastened and copper-bottomed," etc., etc., is long since a thing of the past. The iron age, which succeeded the wooden, has been changed to steel, and steel may change to something else, and steam to electricity. Who knows? Mr. Maginnis, who is himself an engineer and an architect, speaks with authority when he says that, "Whether the improvements be in the ship or in the machinery, gradual advances will be made in the near future." The thirst of competing steamship companies for conquest on the high seas—at any cost —and the ambition of ship-builders to improve upon the latest improvements, will not be satisfied with present attainments, even if it can be proved to a demonstration that thousands of additional horse-power and hundreds of additional tons of coal per day would be required to increase to any appreciable extent the maximum rate of speed that has already been reached. In the meantime some idea may be formed of the possible saving in the consumption of fuel when it is stated that, by a system of induced draught, discovered since the last two Cunarders were designed, the number of boilers necessary to generate steam enough for 30,000 indicated horse-power may be reduced to little more than one-half, which, to put it briefly, means a corresponding saving in space, weight and first cost.[*] In fact, well-informed marine engineers do not hesitate to express their opinion that the day is not far distant when Atlantic greyhounds may be coursing across the ocean at the rate of thirty knots an hour, bringing Queenstown and Sandy Hook within ninety-three hours of each other.

It is difficult to form a correct idea, from any verbal or pictorial representation, of the elegance, the convenience and the comfort attaching to the " Express Steamship." Nothing short of a voyage or voyages in one of these floating palaces would suffice to give an adequate conception of their excellence. And yet, when all is said that can be said in praise of the steamship, some of us "old stagers" can look back, if not with lingering regret, at least with pleasant recollection, to the days of the packet-ship, and even of the sailing vessel of humbler pretensions.

Some of the early emigrant ships were certainly of a mean order, and many emigrants suffered cruel hardships before they reached their destination. It was not an uncommon thing for five or six hundred men, women and children to be huddled together

[*] "The Atlantic Ferry," p. 175.

indiscriminately in the hold of a vessel of from 250 to 300 tons, doomed to subsist on coarsest food, and liable to be immured beneath hatches for days or weeks at a time, without medical attendance, obliged to cook their own food, and scantily supplied with water; and all this for eight or ten weeks at a stretch!

In one of his autobiographic sketches the late Bishop Strachan says that he sailed from Greenock in the end of August, 1799, "under convoy," and such was then the wretched state of navigation, he did not reach Kingston, by way of New York and Montreal, till the 31st of December. In a letter before me an aged friend recites the story of his adventurous voyage from Liverpool to Quebec, some fifty years ago. The ship was a superannuated bluff-bowed East Indiaman, but counted good enough in those days to carry five hundred emigrants across the stormy Atlantic. When ten days out they encountered a hurricane which drove the vessel out of her course. Her three masts fell overboard. The cook's galley and the long boat, the water casks, and everything else on deck, vanished in the gale. The huge hulk rolled like a log in the Bay of Biscay for several days, the passengers meanwhile being confined between decks in horrible confusion. A passing steamer towed them back to Plymouth, where six weeks were spent in refitting the ship, each adult receiving ten shillings and sixpence per week for board and lodging until the repairs were completed. After seven weeks more of great discomfort "and tyrannical treatment on the part of the captain," they finally reached Quebec in 107 days after first embarking at Liverpool.

My own experience of sailing ships, though fifty-seven years have elapsed, is still fresh in mind and recalls some pleasant memories. My first voyage to New York was from the Clyde in a new American ship, commanded by one Captain Theobald, a typical New Englander, as fine a man as one could desire to meet. The voyage was uneventful in the ordinary sense of the term, but one's first voyage in a sailing ship is an event never to be forgotten. It was anticipated with peculiar interest, and regarded with far greater importance than attaches to crossing the Atlantic nowadays. So far from being monotonous, there were incessant changes in sea and sky, in the dress of the ship, and the occupations and songs of the sailors. One day the ship might be bowling along beautifully, decked out in her royals and sky-sails, her studding-sails and stay sails; next day, perhaps, she might be scudding under reefed topsails before an easterly gale, pooping seas that washed the quarter-deck and tumbled like a waterfall into the waist of the ship. Occasionally, a "white squall" coming up would make things lively on deck while it lasted. If becalmed in the right place we caught cod-fish. For the most part, however, the familiar refrain of "tacks and sheets" would be heard many times a day and in the night watches, as we tacked this way and that way against westerly breezes, thankful if the log showed that we had advanced on our course forty or fifty miles in twenty-four hours.

My second voyage westward in a sailing ship was also a memorable one. The Scotch captain of the good ship *Perthshire,* in which we sailed from the Tail of the Bank, off Greenock, on June 19th, 1844, was very unlike the Yankee skipper of the previous voyage. Captain S —— was kind and attentive to his passengers, but not at all popular with his crew. As I watched him taking the sun, the first day out, he said, "Young man, you are going to be some weeks on board this ship, with nothing to do but to eat and drink and sleep. Suppose you take a few lessons in navigation? Here is a spare quadrant which you can use." I jumped at the offer, and very soon mastered at least the outlines of the business. Much was learned in these six weeks—how to find the latitude and longitude at sea; to ascertain the precise deviation of the chronometer from Greenwich time, and of the compass from its true bearing; to measure the trend and velocity of ocean currents, and, failing solar observations, how to consult the moon and the stars. This was not only interesting; it was a fascinating pastime. The captain of a twenty-knot steamship has seldom need to "resolve a traverse;" he steers a straight course for his destination, and can usually estimate within a few hours, or even minutes, when he will reach it. It is quite different with the master of a sailing vessel; after contending with contrary winds and being driven out of his course for weeks at a time, he must often wrack his brains before he can locate his exact position on the chart. To be enveloped in dense fog in the near neighbourhood of Sable Island for several days at a time, as happened to us on this voyage, is a very perplexing position to be in.

For a slight offence Captain S— would send a man aloft to scrape masts in a gale of wind; for a graver misdemeanour he would clap him in irons; had the lash been permitted, he would probably not have hesitated to use it. As might be supposed, things did not go very well in the fo'castle. At length a climax was reached, when the starboard watch came aft one day and lodged a complaint. Getting little or no satisfaction, they retired sullenly, went below, and refused to work for a whole week. The working of the ship then devolved on the first and second mates, the carpenter and the cook, with such of the cabin passengers as could give them assistance. The steerage passengers, siding with the sailors, would not touch a rope, and things even went so far that one of them was placed in confinement for insolence. Some of us were rather glad of the opportunity thus afforded of running up the rigging and creeping through the lubbers' hole without being "salted." When orders were given to shorten sail or shake out a reef, we "lay out" on the yard in sailor fashion; but how much good we did on such occasions will never be known.* At any rate, we counted it fine fun, and it gave the *fiasco* a touch of romance that we slept with loaded pistols under our pillows. But the mutiny ended harmlessly when the pilot came on board. One may cross the Atlantic nowadays without any kind of

---

* If my recollection serves me aright, there were not more than a dozen cabin passengers, and the only one of them who ventured aloft with me was my now venerable friend, Mr. Robert W. Graham, of the Montreal *Star*.

"adventure" like that to adorn a tale, even without so much as once speaking to the captain.

Not every one has the chance of seeing Jack in his citadel. I was deputed by the captain to interview the strikers and endeavour to pacify them. Armed with a copy of the shipping articles which the men had all signed, and another formidable document printed in very large type, I went down into the dingy cabin at the dinner hour. Such a place as it was! I shall never forget it. It corresponded in minute detail to Dana's description of his fo'castle in "Two Years Before the Mast." It was devoid of furniture. There was not even a table to place their food on. In the centre of the floor stood a dirty-looking wooden tub containing a junk of boiled salt beef; near it was a pail full of boiled rice and some hard-tack. The men, about a dozen of them, sat each man on his sea-chest, using his jack-knife to cut and carve with. There were no plates. Imagine the rest. The only grievance they would mention to me was that they had been refused molasses with their rice! Their mind was made up to stay under hatches till the pilot came aboard. They would work for him, but not for the captain; and they kept their word. As I was about leaving, the spokesman of the party, pointing to the mess on the middle of the floor, said with a look that constrained pity, "Mister, how would you like that for your own dinner?" He had the best of the argument. It may be added here that this voyage to New York lasted forty-two days, and the last entry in my log is to the effect that we made as good a passage as any ship from England, "beating the *Columbus* packet-ship by two days!"

The clipper "packet ship" was a vast improvement on the ordinary sailing ship. It had just reached its highest point of development when the ocean steamship first made its appearance. It was to the upper

"GREAT REPUBLIC."
Last of the Clipper Passenger Packets, 1854.

strata of the travelling community, sixty years ago, the counterpart of the express steamer of to-day. The packet-ship was built for fast sailing, with very fine lines, was handsomely fitted up and furnished, was exceedingly well found in eatables and drinkables, and carried a great spread of canvas. To see one of these ships under full sail was a sight to be remembered—a rare sight, inasmuch as all the conditions of wind and water necessary for the display of every stitch of canvas are seldom met with in the North Atlantic. They not unfrequently crossed in fourteen or fifteen days. In winter they might be three months on a single voyage, but their average would be from twenty-five to thirty days.

There were many separate lines of packet-ships sailing at regular intervals from London and Liverpool, and from Hamburg and Havre, to New York, Boston, Philadelphia, and other American ports. Among these were the famous Black Ball Line, the White Star Line, the Old and the New Line of Liverpool packets, etc. The New Line was American, and of it E. K. Collins, the promoter of the Collins' Line of steamers, was the New York agent. The ships were named *Shakespeare, Siddons, Sheridan, Garrick*, and so forth, hence this was called the "Dramatic Line." It is refreshing to read one of their advertisements in the Montreal *Gazette*, as old as November 20th, 1838:

> "These ships are of the first-class, upwards of 800 tons burthen, built in the city of New York, with such improvements as to combine great speed with unusual comfort to passengers. Every care has been taken in the arrangement of their accommodation. The price of passage hence is $140, for which ample stores, including wines, etc., will be provided; without wines, etc., $120. These ships will be commanded by experienced masters, who will make every exertion to give general satisfaction. Letters charged at the rate or 25 cents per single sheet.
>
> ☞ The ships of this line will hereafter go armed, and their peculiar construction gives them security not possessed by any other but vessels of war."
>
> E. K. COLLINS, NEW YORK.
> WM. & JAS. BROWN & CO., LIVERPOOL.

The *Great Republic*, one of the last of the clipper packet-ships, was built in the United States in 1854. She was a four-master of 3,400 tons, 305 feet long, 53 feet beam, and 30 feet in depth. She made the run from New York to the Scilly Islands in thirteen days. She ended her sailing career as a French transport ship, and finally was degraded to a coal hulk. The largest sailing vessel afloat at the present time is the five-masted steel ship *La France*, built on the Clyde by D. & W. Henderson for French owners. She is 6,100 tons burthen, 375 feet long, 49 feet wide and $33\frac{3}{4}$ feet depth. Her fore mainmast is 166 feet high. On her first trip from Cardiff to Rio Janeiro she carried 6,000 tons of coal, and attained a speed of twelve and a half knots.

The Dawn of Steamship Navigation.

Paddle-wheels for driving boats through the water were used long before steam-engines were thought of. They were worked by hand and foot-power without, however, any advantage over the old-fashioned oar. The horse-boat, in a variety of forms, has been in use for many years, and is not yet quite obsolete. In its earlier form two horses, one on each side of a decked scow, were hitched to firmly braced upright posts at which they tugged for all they were worth without ever advancing beyond their noses, but communicating motion to the paddle-wheels by the movable platform on which they trod. For larger boats four or five horses were harnessed to horizontal bars converging towards the centre, and moved around the deck in a circle, the paddles receiving their impulse through a set of cog-wheels. The "latest improvement" was on the direct self-acting treadmill principle, the power being regulated by the weight of the horses and the pitch of elevation given to the revolving platform on which the unfortunate animals were perched. Newcomen's steam-engine had been invented

HORSE-BOAT AT EMPY'S FERRY, OSNABRUCK, ONT.

and used for other purposes eighty years at least, before it was applied to the propelling of vessels. The modern steamboat is not an *invention*, but rather the embodiment of many inventions and experiments, extending over a long series of years by different men and in different countries.

One of the first actual steamboats of which there is authentic record sailed down the River Fulda, in Prussia, in the year 1707. It was built, engined and navigated by a clever Frenchman, Denis Papin,* who was born in 1647, was educated as a physician, and became assistant to the celebrated philosopher, Huygens, in Paris, where he published a small volume on the mechanical effects to be obtained by means of a vacuum. While this attracted the attention of *savants*, it had little or no interest for practical men, and yet in it lay the germ of the power that was to revolutionize the world. He went to London with letters to the Royal Society, and was employed by that society several years, during which he continued his experiments on atmospheric pressure and the vacuum, and the power of steam. He was next appointed Professor of Mathematics in the University of Marburg, from which he removed to Cassel. He had seen the horse-boat in England, and the idea of employing steam to turn the paddles took strong hold of him. He had a boat built and fitted with a steam-engine, in which he embarked with his family and all his belongings, with a view to making his experiment known in Britain and exhibiting his steamboat. All went well until he reached the junction of the rivers Fulda and Weser, where the boatmen got up a hue-and-cry that their craft was endangered by this innovation. In vain Papin protested that he merely wanted to leave the country. On the plea that their rights of navigating these waters had been infringed upon, they rose up *en masse*, seized the steamboat, dragged out the machinery and smashed it to atoms.

* "Denis Papin," by Henry C. Ewart, in *Sunday Magazine*, 1880, p. 316.

Poor Papin found his way back to London a broken-hearted man, never to see the day when his great discovery was to enrich the world.

Fifty years later another experiment was made by

MILLER'S TWIN-BOAT ON LOCH DALSWINTON, 1788.
From "Chambers' Book of Days."

Patrick Miller, a banker in Edinburgh, aided by Mr. Taylor, tutor in his family, and Alexander Symington, a practical engineer. Mr. Miller had a boat built and fitted with a small steam-engine, for his amusement, on Dalswinton Loch, Dumfriesshire. It was a twin-boat, the engine being placed on one side, the boiler on the other, and the paddle-wheel in the centre. It was launched in October, 1788, and attained a speed of five miles an hour. The engine, of one horse-power, is still to be seen in the Andersonian Museum, in Glasgow. Encouraged by his experiment, Mr. Miller bought one of the boats used on the Forth and Clyde Canal, and had a steam-engine constructed for it by the Carron Ironworks

SYMINGTON'S "CHARLOTTE DUNDAS," 1802.
From "Our Ocean Railways."

Company, under Symington's superintendence. On December 26th, 1789, this steamboat towed a heavy load on the canal, at a speed of seven miles an hour; but, strange to say, the experiment was dropped as soon as it was tried.

In 1801 the London newspapers contained the announcement that an experiment had taken place on the Thames, on July 1st, for the purpose of propelling a laden barge, or other craft, against the tide, by means of a steam-engine of a very simple construction. "The moment the engine was set to work the barge was brought about, answering her helm quickly, and she made way against a strong current, at the rate of two and a half miles an hour." In 1802 a new vessel was built expressly for steam navigation, on the Forth and Clyde Canal, under Symington's supervision, the *Charlotte Dundas*, which was minutely inspected on the same day by Robert Fulton, of New York, and Henry Bell, of Glasgow, both of whom took sketches of the machinery to good purpose.* This boat drew a load of seventy

* Mr. Symington's account of his interview with Mr. Fulton, as given in the "Encyclopædia Britannica," is as follows: "When engaged in these experiments, I was called upon by Mr. Fulton, who told me he was lately from North America, and intended returning thither in a few months, but could not think of leaving this country without first waiting upon me in expectation of seeing the boat, and procuring such information regarding it as I might be pleased to communicate. . . . In compliance with his earnest request, I caused the engine fire to be lighted up, and in a short time thereafter put the steamboat in motion, and carried him four miles west on the canal, returning to the point from which we started in one hour and twenty minutes (being at the rate of six miles an hour), to the great astonishment of Mr. Fulton and several gentlemen, who at our outset chanced to come on board. During the trip Mr. Fulton asked if I had any objection to his taking notes regarding the steamboat, to which I made no objection, as I considered the more publicity that was given to any discovery intended for the general good, so much the better. . . . In consequence he pulled out a memorandum book, and, after putting several pointed questions respecting the general construction and effect of the machine, which I answered in a most explicit manner, he jotted down particularly everything then described, with his own observations upon the boat during the trip."

tons, at a speed of three and a half miles an hour, against a strong gale of wind. Under ordinary conditions she made six miles an hour, but her admitted success was cut short by the Canal Trust, who alleged that the wash of the steamer would destroy the embankment.

BELL'S "COMET." *

Nothing more was heard of the steamboat in Britain until 1812, when Henry Bell surprised the natives of Strathclyde by the following advertisement in the Greenock *Advertiser*:

STEAM PASSAGE BOAT,

"THE COMET,"

BETWEEN GLASGOW, GREENOCK AND HELENSBURGH,

FOR PASSENGERS ONLY.

The subscriber having, at much expense, fitted up a handsome vessel, to ply upon the River Clyde, between Glasgow and Greenock, to sail by the power of wind, air and steam, he intends that the vessel shall leave the Broomielaw on Tuesdays, Thursdays and Saturdays, about mid-day, or at such hour thereafter as may answer from the state of the tide; and to leave Greenock on Mondays, Wednesdays and Fridays, in the morning, to suit the tide.

The elegance, comfort, safety and speed of this vessel requires only to be proved to meet the approbation of the public; and the proprietor is determined to do everything in his power to merit public encouragement.

The terms are, for the present, fixed at 4s. for the best cabin, and 3s. for the second; but beyond these rates nothing is to be allowed to servants, or any other person employed about the vessel.

The subscriber continues his establishment at HELENSBURGH BATHS, the same as for years past, and a vessel will be in readiness to convey passengers to the *Comet* from Greenock to Helensburgh.

HENRY BELL.

HELENSBURGH BATHS, *5th August, 1812.*

Bell's *Comet* was a quaint-looking craft, with a tall, slender funnel, that served the double purpose of mast and chimney. Her length was 42 feet, breadth

* "The Story of Helensburgh," 1894, p. 92.

11 feet, draught of water 5½ feet. She had originally two small paddle-wheels on each side with four arms to each. The engine was about three horse-power, and seems to have been the joint production of Bell and the village blacksmith. The boiler was made by David Napier, at a cost of £52. The engine is still

BELL'S "COMET," OFF DUMBARTON ON THE CLYDE, 1812.
From "Chambers' Book of Days."

preserved in the patent office of the South Kensington Museum. The *Comet* was lengthened at Helensburgh, in 1818, to 60 feet, and received a new engine of six horse-power, by means of which her speed was increased to six miles an hour. This engine was made by John Robertson, of Glasgow.

The *Comet* did not pay as a passenger boat on the

Clyde, and was soon after her launch put on the route to Fort William, and continued on that stormy route till December 15th, 1820, when she was wrecked at Craignish, on the West Highland coast. She had left Oban that morning against the advice of her captain, who deemed the boat unseaworthy and quite unfit to encounter the blinding snow-storm, in the midst of which she went ashore. But Bell had overruled the captain. Fortunately there was no loss of life. She was replaced in the following year by a larger and improved style of vessel, called by the same name and sailed by the same master, Robert Bain, who was the first to take a steamer through the Crinan Canal, and the first to traverse the Caledonian Canal from sea to sea by steam, in 1822. The second *Comet* came into collision with the steamer *Ayr* off Gourock in October, 1825, and sank with the loss of seventy lives. She was raised, however, was rigged as a schooner, renamed the *Anne*, and sailed for many years as a coaster.

Mr. Bell was born in Linlithgow in 1767. The son of a mechanic, he worked for some time as a stonemason, afterwards as a carpenter, and gained some experience in ship-building at Bo'ness under Mr. Rennie. He removed to Helensburgh in 1808, where his wife kept the Baths Inn while he was experimenting in mechanical projects. He was a man of energy and enterprise, but like most inventors was always scant of cash. Had it not been for the generosity of his friends, and an annuity of £100 which he received from the Clyde Trust, he would have come to want in his old age. He seems to have had steam navigation on the brain as early as 1786, and had communicated his ideas on the subject to most of the crowned heads of Europe, as well as to the President of the United States, before he built the *Comet*. Mr. Bell's memory is perpetuated in an obelisk erected by the city of Glasgow corporation on a picturesque promontory on the banks of the Clyde at Bowling, " in acknowledgment of a debt which it can never repay." There is also a handsome granite obelisk to his memory on the esplanade at Helensburgh, the inscription on which testifies that " Henry Bell was the first in Great Britain who was successful in practically applying steam power for the purpose of navigation." The stone effigy of the man adjoining his grave in Row churchyard was placed there by his friend Robert Napier, whose fame and fortune were largely the result of Bell's enterprise. Mr. Bell died at his inn in Helensburgh, November 14th, 1830.

Fifty years later witnessed the full development of Mr. Bell's ideal in the *Columba*, then as now the largest river steamer ever seen on the Clyde, and the swiftest. The *Columba* is built of steel, is 316 feet long and 50 feet wide. She has two oscillating engines of 220 horse-power, and attains a speed of twenty-two miles an hour. Her route is from Glasgow to Ardrishaig and back, daily in summer, when she carries from 2,000 to 3,000 persons through some of the finest scenery in Scotland. She is provided with steam machinery for steering and warping her into the piers, and with other modern appliances that make her as handy as a steam yacht. She resembles a little floating town, with shops and post-office where you can procure money orders and despatch telegrams. And what is the *Columba* after all but an enlarged and perfected reproduction of Bell's *Comet!*

The reputation of the Clyde in respect of ocean steamships and "ironclads" has become world-wide. Some of the best specimens of marine architecture

"COLUMBA," FAMOUS CLYDE RIVER STEAMER, 1875.

are Clyde-built. Her own river steamers are the finest and fleetest in the United Kingdom. The Thames river steamers, though far inferior to the Clyde boats, answer their purpose by conveying vast numbers of people short distances at a cheap rate. The Victoria Steamboat Association, with its fleet of forty-five river steamers, can carry 200,000 people daily for a penny a mile. The Rhine steamers and those plying on the Swiss lakes are in keeping with the picturesque scenery through which they run.

Painted in bright colours, they present a very attractive and smart appearance. They are kept scrupulously clean and are admirably managed. Many of them are large, with saloon cabins the whole length of the vessel, over which is the promenade deck covered with gay awnings. They run fast. The captain sits in state in his easy chair under a canopy

"WILHELM KAISER" ON THE RHINE, 1886.

on the bridge—smoking his cigar. The chief steward, next to the captain by far the most important personage on board, moves about all day long in full evening dress—his main concern being to know what wine you will have for lunch or dinner that he may put it on ice for you. The *table d'hote* is the crowning event of the day on board a Rhine steamer, *i.e.*, for the misguided majority of tourists to whom a swell dinner offers greater attractions than the finest scenery imaginable.

The success of the first *Comet* induced others to follow the example. The year 1814 saw two other small steamboats on the Clyde. Next year the *Marjery*, built by Denny of Dumbarton, made a voyage to Dublin and thence to the Thames, where she plied between London and Margate for some time, to the consternation of the Thames watermen. In 1818 David Napier of Glasgow went into the business, and equipped a number of coasting steamers with improved machinery. At this time the *Rob Roy*, claimed to be the pioneer of sea-going steamers, began to run to Belfast, but being found too small for the traffic she was put on the Dover and Calais route. In 1819 the Admiralty of the day had a steamboat built for towing men-of-war, called the *Comet*, 115 feet by 21 feet, with two of Boulton & Watt's engines of 40 horse-power each. This vessel was followed by the *Lightning, Echo, Confiance, Columbia* and *Dee*—the latter vessel having side-lever engines of 240 horse-power, with flue boilers carrying a pressure of six pounds to the square inch, which developed a speed of seven knots an hour. In 1822 a large number of steam vessels fitted with condensing engines were afloat. The *James Watt* was built in that year to ply between Leith and London. The largest steamer at that time was the *United Kingdom*, built by Steele of Greenock, 160 feet long by 26½ feet wide, having engines of 200 horse-power—as much an object of wonder in those days for her "gigantic proportions" as was the

Great Eastern thirty years later. In 1825 there were 168 steam vessels in Britain; in 1835 there were 538; in 1855 there were 2,310, including war vessels afloat and building; in 1895 the number of steam vessels built in the United Kingdom was 638, of which number 90 per cent. were built of steel. In 1897 the number of steamers over 100 tons in the United Kingdom, including the colonies, was computed to be 8,500, with a net tonnage of 6,500,000 tons.

THE "CLERMONT."

Three years before Bell's achievement on the Clyde, a clever American, profiting by the experiments of Symington, applied his inventive genius to perfecting the application of steam as a motive power for vessels, and gained for himself the honour of being the first to make it available for practical use on a paying basis. This was Robert Fulton, a native of Pennsylvania, born in 1765, who commenced business as a portrait painter and followed that profession for some years in France and England. He invented a number of "notions," among the rest a submarine torpedo-boat, in which he claimed that he could remain under water for an hour and a half at a time; but failing to receive the patronage of any naval authorities, he returned to New York, and, with the assistance of Mr. John Livingstone, had a steamboat built and fitted with an English engine by Boulton & Watt, of Birmingham. The *Clermont* (after being lengthened) was 133 feet long, 18 feet beam, and $7\frac{1}{2}$ feet deep. Her wheels were uncovered, 15 feet in diameter, with eight buckets, 4 feet long, to each wheel, and dipping 2 feet. The cylinder was 24 inches in diameter, with 4 feet stroke of piston. The boiler was of copper, 20 feet long, 7 feet wide and 8 feet high.

The *Clermont* made her first voyage from New York to Albany, August 7th, 1807. Her speed was

FULTON'S "CLERMONT" ON THE HUDSON, 1807.

about five miles an hour. During the winter of 1807-8 she was enlarged, her name being then changed to *North River*. She continued to ply successfully on the Hudson as a passenger boat for a number of years, her owners having acquired the exclusive right to navigate the waters of the State of New York by steam. The *Car of Neptune* and the *Paragon*, of 300 and 350 tons, respectively, were soon added to the Fulton & Livingstone Line. Both of these vessels were fitted with English engines. The *Paragon* continued to ply on the Hudson for about ten years, earning a good deal of money for the owners. About 1820, while ascending the river, she ran upon a rock and became a total wreck. Other steamboats were built for other waters, and very soon there were steamers plying on all the navigable rivers of the

MISSISSIPPI STEAMBOAT "J. M. WHITE," 1878.

United States available for commerce. Mr. Fulton married a daughter of Mr. Livingstone. He died in New York in 1815, at the height of his fame and prosperity.

The contrast between Fulton's *Clermont,* or Bell's *Comet* and the Atlantic Liner coursing over the sea at railway speed is very striking, and scarcely less remarkable the comparison of the river steamboat of to-day with these early experiments. America has developed a type of steamboat, or rather types of steamboats, peculiarly its own. The light-draught Mississippi steamers* bear little resemblance to the Hudson River and Long Island Sound boats, while the American steam ferry-boat is a thing certainly not of beauty, but unique. Dickens in his American Notes speaks of the *Burlington*, the crack steamer on Lake Champlain in the early forties, as " a perfectly exquisite achievement of neatness, elegance and order —a model of graceful comfort and beautiful contrivance." But Dickens never saw the *Priscilla*. She was only launched in 1894, and is claimed to be " pre-eminently the world's greatest inland steamer— the largest, finest and most elaborately furnished steamboat of her class to be found anywhere." The *Priscilla* is 440½ feet long, 52½ feet wide, or 95 feet over the paddle-boxes. The paddle-wheels are of the feathering type, 35 feet in diameter and 14 feet face. Her light draught is 12½ feet, and her speed easily 22 miles an hour, though the ordinary service of the line does not demand such fast running. Her night's work is 181 miles, which she covers leisurely in ten hours. She cost $1,500,000. All the interior

---

*These cuts, copied from Stanton's "American Steam Vessels," represent first-class Mississippi and Ohio light-draught, high-pressure river steamers. The *J. M. White*, of 1878, was deemed "a crowning effort in steamboat architecture in the West." She was 320 feet long and 91 feet in width, over the guards. Her saloons were magnificently furnished, and all her internal fittings of the most elaborate description. She carried 7,000 bales of cotton and had accommodation for 350 cabin passengers. Her cost was $300,000. She was totally destroyed by fire in 1886.

decorations are very elaborate and handsome. In her triple row of staterooms there is luxurious sleeping accommodation for 1,500 passengers. In the spacious dining-room 325 persons may be seated at one time. The grand saloon is a magnificent spectacle, large and lofty, superbly decorated and lighted by electricity. The *Priscilla* has cargo capacity for 800 tons of

OHIO STEAMBOAT "IRON QUEEN," 1882.

freight. "Her machinery is not only a marvel of design and workmanship, but it fascinates all persons interested in mechanical devices." It consists of a double inclined compound engine, with two high-pressure cylinders, each fifty-one inches in diameter, and two low pressure, each ninety-five inches in diameter, all with a stroke of eleven feet. There are ten return tubular boilers of the Scotch type, each fourteen feet in diameter and fourteen feet long, constructed for a working pressure of 150 lbs. to the square inch. The indicated horse-power is 8,500. The machinery is principally below the main deck, leaving

"PRISCILLA."
Fall River and Long Island Sound Line, 1894.

all the space on and above this deck available for general purposes.

This floating palace was built at Chester, Pa., by the Delaware Iron Shipbuilding and Engine Works Company. She is built of steel. Her registered tonnage is

5,398 tons. Although so vast in her proportions, the *Priscilla* sits on the water as lightly and gracefully as a swan. Painted white as snow outside, as nearly all American river steamers are, she presents a beautiful, you might say a dazzling, appearance; and she is only one of five magnificent steamers of the Fall

"NEW YORK."
The latest Hudson River Day Steamer, 1897.

River Line, all substantially alike in design and equipment, running regularly all the year round between Fall River and New York, with a perfection of service that cannot be surpassed.

This cut, kindly furnished by the owners, gives a faithful representation of the exterior of a very beautiful Hudson River day steamboat. The *New*

York is built of steel, 311 feet over all, breadth of beam 40 feet, and over the guards 74 feet; average draught of water 6 feet. She combines speed, luxuriousness of furnishing and a beauty of finish in all parts that has not been surpassed on vessels of this class. She is capable of running 24 miles an hour. This boat and her consort, the *Albany*, are claimed to be the finest day passenger river steamers in the world. She is not crowded with 2,500 passengers, of whom 120 may sit down together to an exquisite dinner in the richly decorated dining-room.

A distinct class of steamboats peculiar to America is the ferry-boat. In one of its forms it is to be found fully developed in New York harbour, and serves to convey daily countless thousands of people whose business lies in New York City, but whose homes are on Brooklyn Heights or elsewhere on Long Island, or the New Jersey coast. The boats are very large and very ugly, but do their work admirably, being adapted for the transport of wheeled carriages of every description as well as for foot-passengers. One of the sights of New York worth seeing is a visit to the Fulton Ferry in the morning or in the evening, when the crowds are the greatest. The *Robert Garrett*, which runs down the bay to Staten Island, carries from 4,000 to 5,000 passengers at a trip, and is said to be the largest steam-ferry passenger boat in existence. She is owned by the Staten Island Rapid Transit Co., and cost $225,000.

Another type of ferry-boat is that which, in addition to carrying passengers, is specially adapted for railway purposes. The best specimen of this kind of steamboat is probably to be found on Lake Erie, where a pair of boats, precisely alike, keep up regular communication twice a day, summer and winter, between Coneant, Ohio, and Port Dover, Ontario. They are named *Shenango*, 1st and 2nd. They are each 300 feet long and 53 feet in width. On the main deck are four railway tracks, sufficient for twenty-six loaded cars

"ROBERT GARRETT," FERRY STEAMBOAT, NEW YORK.

each containing 60,000 lbs. of coal. On the upper deck are handsomely fitted cabins for 1,000 passengers. The ferry is sixty-five miles wide. Sometimes it is pretty rough sailing; but these steamers never fail to make the round trip in thirteen hours. They are fitted with compound engines, Scotch boilers, and twin screws; they draw $12\frac{1}{2}$ feet of water when loaded and run twelve miles an hour; they are prodigiously strong, and can plough their way through fields of ice with marvellous facility.

CHAPTER II.

EARLY YEARS OF STEAM NAVIGATION.

The *Accommodation*—The *Savannah*—*Enterprise*—*Royal William*—*Liverpool*—*Sirius* and *Great Western*—*Great Britain* and *Great Eastern*—The Brunels—The screw propeller.

TWO years after the *Clermont* had commenced to ply on the Hudson, and three years before the *Comet* had disturbed the waters of the Clyde, the first steamboat appeared on the St. Lawrence. The *Accommodation*, built by the Hon. John Molson, of Montreal, made her maiden trip to Quebec on November 3rd, 1809, carrying ten passengers, in thirty-six hours' running time. In accordance with the usual custom, which continued for many years, she anchored at night, so that the whole time occupied in the voyage was sixty-six hours. If she ascended the St. Mary's current, she was towed up by oxen. The length of this vessel was eighty-five feet over all, her breadth sixteen feet, her engine was of six horse-power, and her speed five miles an hour. The *Accommodation* was built at the back of the Molson's Brewery, and was launched broadside on. Her engine was made by Boulton & Watt, of Birmingham, England. The fare from Montreal to Quebec by this vessel was £2 10s.; children, half price; "servants with *birth* (sic), £1 13s. 4d.; without *birth*, £1 5s." The Quebec *Mercury*, announcing her arrival, remarked: "She is incessantly crowded with visitors. This steamboat receives her impulse from an open-spoked perpendicular wheel on each side, without any circular band or rim. To the end of each double spoke is fixed a square board which enters the water, and by the rotatory motion of the

"JEANIE DEANS," CLYDE STEAMBOAT.
From "Mountain, Moor and Loch," London, 1894.

wheels acts like a paddle. No wind or tide can stop her."

The Savannah.—In the year 1818 there was built in New York, by Messrs. Crocker and Pickett, a full-rigged sailing ship of about 350 tons, named the *Savannah*. She was intended to be used as a sailing packet between New York and Havre, but before she was completed she was purchased by William Scarborough & Co., a shipping firm in Savannah, who fitted her up with a steam-engine of 90 horse-power, placed on deck, and a pair of paddle-wheels enclosed with canvas coverings, so constructed that they could be folded up and taken on deck in stormy weather, and that tedious operation seems to have been gone through pretty frequently in the course of her first voyages. Her maiden trip from New York to Savannah occupied 8 days, 15 hours. She left Savannah for Liverpool under steam, May 22nd, 1819, and arrived in the Mersey, " with all sail set," on June 20th, making the run in twenty-nine and a half days. The whole time that the engine was at work during the voyage is said to have been only eighty hours. " She hove to off the bar, waiting for the tide to rise, at 5 p.m. shipped her wheels "—so the record of the period runs—" furled her sails and steamed up the river, with American banners flying, the docks being lined with thousands of people, who greeted her arrival with cheers." From Liverpool, the *Savannah* sailed up the Baltic to Stockholm and St. Petersburg. On her return voyage, on account of stormy weather, the engine was scarcely used at all until the pilot came aboard off Savannah, when the sails were furled, and with the flood-tide she steamed into port. After several voyages of a similar kind, the machinery was removed and she plied for some time as a sailing packet between New York and Savannah, and was eventually wrecked on Long Island in 1822.

Shortly after this the British Government offered a prize of £10,000 to the party who should first make a successful voyage by steam power to India. The prize was won by Captain Johnston, who sailed from England on August 16th, 1825, in the *Enterprise*, of 500 tons and 240 horse-power,* and reached Calcutta on the 7th of December. The distance run was 13,700 miles, and the time occupied 113 days, during ten of which the ship was at anchor. She ran under steam

THE "SAVANNAH," 1819.

sixty-four days and consumed 580 chaldrons of coal, the rest of the voyage being under sail.

Eight years followed without any further attempts in the direction of ocean steam navigation. There seemed to be nothing in these costly experiments that would induce capitalists to invest their money in steamships. Sailing vessels had crossed the Atlantic in much less than thirty days, and had made the

* "Our Ocean Railways," p. 69.

voyage to India in less time than the *Enterprise* took to do it. It would not pay! and had not scientific men and practical engineers pronounced the idea of transatlantic steamships as Utopian and utterly impracticable? "No vessel could be constructed," they said, "that could carry enough coal to take her across the Atlantic by steam power alone." Some of these unbelievers lived to see the day when large ocean steamers not only carry enough coal to take them from Liverpool to New York, but actually enough for the return voyage also.

The "Royal William."

The *Savannah* and *Enterprise* were admittedly nothing more than sailing ships with auxiliary steam power. In the archives of the National Museum at Washington there is to be found the full history and log of the *Savannah*, which proves conclusively that she was not entitled to be called the pioneer of transatlantic steam navigation. That the honour belongs to the *Royal William*, built at Quebec and engined at Montreal, has been clearly proven. The evidence, in support of this claim is embodied in a report of the Secretary of State of Canada for the year ended December 31st, 1894. From this it appears that the *Royal William* was designed by Mr. James Goudie, Marine Architect of Quebec, and that she was launched from the shipyard of Messrs. Campbell and Black at Cape Cove, Quebec, April 29th, 1831, in presence of Lord Aylmer, the Governor-General,

MODEL OF STEAMSHIP "ROYAL WILLIAM."

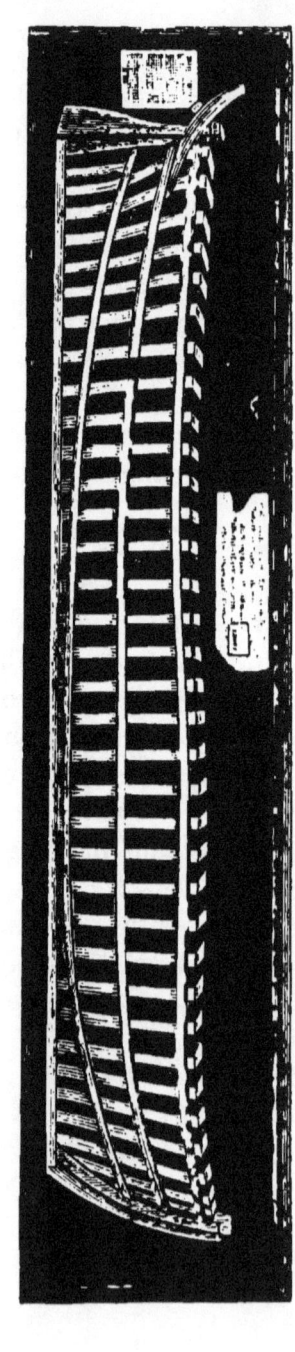

THIS INTERESTING RELIC HAS AN HONOURED RESTING-PLACE IN THE LIBRARY OF THE LITERARY AND HISTORICAL SOCIETY OF QUEBEC. IT WAS SENT, AT THE REQUEST OF THE COMMITTEE OF THE ROYAL NAVAL EXHIBITION, TO THAT EXHIBITION, HELD IN LONDON IN 1891, AND NUMBERED 4,736, WHERE IT ATTRACTED CONSIDERABLE ATTENTION, AND THE SOCIETY RECEIVED FROM THE COMMITTEE A HANDSOME DIPLOMA BY WAY OF A SOUVENIR.

THE IMPORTANCE OF THIS MODEL WAS RECOGNIZED BY THE DOMINION GOVERNMENT, BY ORDERING A FAC-SIMILE OF IT TO BE MADE, AND SENDING IT TO THE COLUMBIAN EXHIBITION, OR WORLD'S FAIR, AT CHICAGO, IN 1893. IT IS NOW TO BE SEEN IN THE DEPARTMENT OF AGRICULTURE AT OTTAWA.

and a vast concourse of people, Lady Aylmer naming the vessel with the usual ceremonies after the reigning monarch, William IV. She was towed to Montreal, where her engines of 200 horse-power were fitted by Messrs. Bennett and Henderson. She steamed back to Quebec in the beginning of August. She was built for the Quebec and Halifax Steam Navigation Company, incorporated by Act of Parliament, March 31st, 1831. This company comprised 235 persons whose names appear in the Act, among them being the three brothers, Samuel, Henry and Joseph Cunard. Samuel, the founder of the Cunard Line, was a frequent visitor at the Quebec shipyard, and carefully noted down all the information he could get from the builders.

This historic vessel was registered No. 2 in the port of Quebec. She was rigged as a three-masted schooner, of $363\frac{60}{84}$ tons burthen, with a standing bowsprit and square stern. Her length was 160 feet; breadth, taken above the main wales, 44 feet; depth of hold, 17 feet 9 inches; and width, between the paddle-boxes, 28 feet. She cost about £16,000. The *Royal William*, commanded by Captain J. Jones, R.N., sailed from Quebec for Halifax, August 24th, 1831, with twenty cabin passengers, seventy steerage, and a good freight. She arrived on the 31st—six and a half days from Quebec. Several voyages were made that year to Halifax and the Gulf ports. Next year, owing to the prevalence of cholera, trade was at a standstill, and there was nothing for the new steamship to do. She was accordingly sold by Sheriff Gugy, at the church door, in the parish of

Sorel, for £5,000. In April, 1833, she was placed under the command of Captain John Macdougall, a native of Oban, Scotland. During May she towed vessels from Grosse Isle, and in June sailed for the lower ports, Halifax and Boston, reaching the latter place on the 17th—the first British steamer to enter that port. On her return to Quebec, her owners decided to send her to London to be sold. She sailed August 5th, arrived at Pictou on the 8th, and sailed thence on the 18th, with seven passengers, a box of stuffed birds, one box and one trunk, some household furniture, 254 chaldrons of coal, and a crew of thirty-six men. The voyage to Cowes, Isle of Wight, was made in nineteen and a half days. She was deeply laden with her coal, had very rough weather, and had to run with one engine for ten days. A short time having been spent at Cowes, painting the ship, etc., "she steamed up to Gravesend in fine style—the first vessel to cross the Atlantic propelled by the motive power of steam alone."

The *Royal William* was sold in London for £10,000, and was chartered to the Portuguese Government as a transport. In 1834 she was sold to the Spanish Government, and named the *Isabel Segunda*, and while in this service was the first war-steamer to fire a hostile shot. In 1837 she was sent to Bordeaux, France, for repairs, but, her timbers being badly decayed, her machinery was transferred to a new vessel of the same name, while she herself terminated her brilliant career as a hulk.*

* Sufficient importance was attached to this matter to cause the two Houses of Parliament, in Ottawa, to order a brass tablet,

Another steamer bearing the name *Royal William* was despatched from Liverpool to New York, by the Transatlantic Steamship Company, in 1838. This was a vessel of 617 tons, and 276 horse-power—the first to make the westward voyage from Liverpool, and the first passenger steamer to cross the sea. After a few voyages of doubtful success, this steamer was degraded into a coal-hulk, and a much larger and faster vessel took her place. This was the *Liverpool*—built expressly for the Atlantic trade, with luxurious fittings for seventy or eighty first-class passengers. She was a fine ship, of 1,150 tons burthen, and 468 horse-power. She sailed from Liverpool, October 20th, 1838, but had to put back to Queenstown on the 30th; sailing thence on November 6th, she reached New York on the 23rd. After several voyages, averaging seventeen days out and fifteen days home, she was sold to the Peninsular and Oriental Company, and was finally wrecked off Cape Finisterre in 1846.

In 1839 the late Sir Hugh Allan and several other Canadians made an adventurous voyage in the *Liverpool*. Sailing from New York, December 4th, they had a succession of gales up to the 28th, when they were scarcely half-way across the Atlantic. The chief

commemorative of the event, to be placed in the corridor of the Library of Parliament. The tablet, of which a fac-simile is presented in our frontispiece, was unveiled with fitting ceremony by His Excellency the Governor-General, on the occasion of the opening of the Colonial Conference, June 28th, 1894.—*Vide:* "The Journals of the Colonial Conference" (*Appendix*) ; "Journal of the House of Commons," 1894 ; " Transactions of the Royal Society of Canada."

engineer then reported that unless things mended they would run short of coal. The chief steward at the same time expressed grave doubts as to his provisions holding out. A consultation having been held, it was resolved to change their course for the Azores. They reached Fayal just as the last shovelful of coal was thrown on the fires. Four days were spent on the Island, during which time the passengers

THE "SIRIUS," 1838.

were treated to a round of festivities. On arriving at Liverpool, they learned that the ship had been given up as lost—not having been heard of since she sailed from New York thirty-nine days before.

The "Sirius" and "Great Western."

The departure of these steamships from England to America in 1838 marks an important epoch in the history of steam navigation, inasmuch as the practicability of establishing a regular transatlantic steam service was now for the first time to be clearly demonstrated. As the *Sirius* made only one round voyage, there is little to be said about her beyond admiring the pluck of her owners. She was a small vessel of about 700 tons and 320 horse-power, built at Leith for the St. George Steam-packet Company, and had plied successfully for some time between London and Cork. She was chartered by the then newly formed "British and American Steam Navigation Company," of which the famous ship-builder, Laird, of Birkenhead, was the leading spirit. The *Sirius* was despatched from London for New York, *via* Cork, whence she sailed on April 4th, with ninety-four passengers. She arrived in New York on the 22nd, after a successful voyage of seventeen clear days, being commanded by Lieut. Roberts, R.N., who was afterwards lost at sea with the ill-fated SS. *President*, in 1841. The return voyage was made in about the same number of days as the outward trip.

The *Great Western*, designed and built by Mr. William Patterson at Bristol, for the Great Western Steamship Company, sailed from Bristol, April 8th, 1838, in command of Lieut. James Hoskin, R.N., and reached New York on the 23rd, making the run in fifteen days with a consumption of 655 tons of coal and realizing an average speed of a little over eight knots an hour. She returned to Bristol in somewhat less than fifteen days. A fine ship she was, of 1,340 tons and 440 horse-power, 212 feet long, and $35\frac{1}{2}$ feet beam. Her best run between New York and Bristol was made in 12½ days,* a remarkable record for that time. Altogether she was admitted to be a distinct success. She was sold in 1847 for £25,000, after which she sailed regularly for ten years to the West Indies. In the meantime the owners of the *Sirius* had built a much larger boat, the *British Queen*, which made her maiden voyage from Portsmouth in 1839. After making a number of voyages to New York this fine ship was sold to the Belgians in 1841, chiefly owing to the collapse of the company occasioned by the loss of a sister-ship, the *President*, which sailed from New York, March 11th of that year, and was never afterwards heard of.

The "Great Britain" and "Great Eastern."

The *Great Britain*, designed by Brunel, and built at Bristol by Mr. Patterson, was the first iron steamship of large dimensions. She was very large for her time, being 322 feet long, 48 feet wide, and 31½ feet deep; her tonnage was 3,270 tons, and her engines 1,500 horse-power. As originally rigged she had six masts; she had a six-bladed screw-propeller, 15½ feet in diameter, which made 18 revolutions per minute, giving her a maximum speed of twelve knots an hour. A very handsome model, of prodigious strength, and a fine sea-boat was the *Great Britain*. She commenced plying to New York, July 26th, 1845, and was a pronounced success. On the 22nd of September, 1846, on her outward voyage, she was stranded on the

* Others say 10½ days.

Irish coast, and became deeply embedded in the sands of Dundrum Bay, where she lay all winter, exposed to violent storms; but she withstood the strain, was raised from her watery grave, was refitted and placed on the Australian route, where she sailed successfully until 1882, when her machinery was taken out and she closed her remarkable career as a full-

THE "GREAT BRITAIN," 1845.

rigged sailing ship, when nearly fifty years old! and was finally used as a coal-hulk at the Falkland Islands, where her remains are still to be seen.

The Great Eastern.—The British Government having in 1853 advertised for tenders to carry the mails to India and Australia, a number of wealthy and scientific men formed themselves into a company called the Eastern Steam Navigation Company, with a capital of £1,200,000, and sent in a tender, but it was not accepted.* The company, however, resolved to build a fleet of steamers, of which the *Great Eastern* was to be the first. Mr. Brunel, who had designed the *Great Britain*, was selected as the architect, and Mr. Scott Russell, as the builder of the pioneer ship. The proposal suited Mr. Brunel's

THE "GREAT EASTERN," 1857.

sanguine temperament, and he recommended the building of a monster iron steamship, that should eclipse all previous efforts in marine architecture, a vessel that should run, say, to Ceylon at an average speed of fifteen knots, and carry coal enough to take her out and home again. From Ceylon smaller boats would continue the service to India and Australia. The embodiment of Mr. Brunel's magnificent concep-

* Fry's "History of Steam Navigation," p. 182.

tion was the *Great Eastern*, skilfully wrought out, but destined to prove a gigantic failure.

This extraordinary ship was commenced at Millwall on the Thames, in May, 1854, and was completed in 1857, at a cost of nearly $5,000,000. When ready for launching, her estimated weight was some 12,000 tons. As no such load had ever before slid down the ways of a shipyard, every precaution and appliance that skill could suggest were brought into requisition. She was to be hauled down, broadside on, by an elaborate arrangement of chains and stationary engines; but when the critical moment arrived the ponderous mammoth would not budge, and it cost something like $600,000 and constant labour for three months before she reached her destined element. The *Great Eastern* was 692 feet long, 83 feet in width, and $58\frac{1}{2}$ feet deep. She was reckoned at 22,500 tons burthen. Her four engines were collectively of 11,000 indicated horse-power. She was fitted up in grand style to accommodate 4,800 passengers. As a troop-ship she could carry comfortably an army of 10,000 men in addition to her own crew of 400. She was provided with both paddle-wheels and a screw-propeller. The wheels were fifty feet in diameter, making twelve revolutions per minute; the four-bladed screw was twenty-four feet in diameter, adapted for forty-five revolutions per minute. Her estimated speed was fifteen knots, but her best average never exceeded twelve knots. Her first voyage from Southampton to New York was made in 10 days and 21 hours; the highest speed by the log was fourteen and a half knots, and the greatest day's run three hundred and thirty-three knots. Her arrival in New York, June 27th, 1860, created a great sensation. Fort Hamilton saluted her with a discharge of fourteen guns—the first instance of a merchant vessel being thus honoured in America. She returned home *via* Halifax, making the run thence to Milford Haven in 10 days and 4 hours. In May, 1861, she made another voyage to New York, carrying one hundred passengers, but with no improvement in her speed. On her return to Liverpool she was chartered by the British Government to bring out troops to Canada. She arrived at Quebec, July 6th, 1861, with 2,528 soldiers and forty civilians, and during her stay there was visited by large crowds of people. Leaving Quebec, August 6th, she reached Liverpool on the 15th. A couple more voyages to New York, and her career as a passenger ship was ended. She had been singularly unfortunate. Her first commander, Captain Harrison, was drowned in the Solent by the upsetting of a small boat. On her trial trip, by the bursting of a steam jacket, six of her crew were killed and the ship was badly damaged. She had broken her rudder in mid-ocean, and lay for days a helpless mass in the trough of the sea during a gale of wind, rolling frightfully. Worse than all, she had got on the rocks entering New York harbour, with serious damage to her hull. The momentous question arose, What was to be done with her?

This leviathan of the deep was finally fitted up as

a "cable ship," and for a short time did good service in that line. In 1865 she had laid the second Atlantic cable to within a few hundred miles of Newfoundland, when it snapped and disappeared in 1,950 fathoms of water. Next year the *Great Eastern* not only was the means of laying a new cable successfully, but was the means of picking up the lost one—a remarkable feat of seamanship and electrical skill. After laying several other cables the big ship was tied up, never to go again. She was eventually sold for £16,000 and broken up, a somewhat tragic ending for such a triumph of engineering skill. But who can tell how much the successful "liner" of to-day owes to the failure of the *Great Eastern?* She came out ahead of time, and when the intricate art of managing successfully the details of an ocean steamship had yet to be learned.

Isambard Kingdom Brunel, born at Portsmouth in 1806, was the son of Sir Mark I. Brunel, a French engineer, who attained celebrity as the architect of the Thames Tunnel, and other important works, in which he was assisted by his son, who also became famous as the Engineer-in-Chief of the Great Western Railroad, in the construction of which he adopted the broad gauge (7 feet), against the remonstrances of Stephenson and other railway authorities, and which was eventually changed to what has become the national gauge (4 feet, $8\frac{1}{2}$ inches), at enormous expense. Mr. Brunel died in 1859. It was his misfortune to have landed on this planet about fifty years too soon.

The Screw-Propeller.

Most people fail to find much resemblance, if any at all, between that comparatively small-looking two or three-bladed thing that drives the steamship through the water at the rate of twenty miles an hour, and what is commonly known as a screw; but the discrepancy is easy of explanation. Archimedes, who is credited with the invention of the screw as a mechanical lever, little dreamed of the uses to which it was to be turned two thousand years later. He is said to have employed the screw in launching a large ship, pushing it into the water as is now done by hydraulic appliances. By changing his fulcrum and making the screw a part of the ship, the modern engineer has only reversed the mode of applying propelling power; the principle is the same. The effect produced by the screw in propelling a ship will be best understood by supposing an ordinary screw of large dimensions to be revolving rapidly in a trough full of water. It would then send the water away from it with great force; but as action and reaction are equal it would be itself, at the same time, urged in the opposite direction with exactly the same degree of force. If we suppose it, then, to be fixed in a ship, the ship will be pushed forward with the same force that is exerted by the screw in pushing back against the water. If the screw is made to revolve in the opposite direction, the converse of this takes place, and the ship is pushed backwards by the reaction of the screw.* The idea

---

* Encyclopedia Britannica, 8th Ed., Vol. xx, p. 657.

has long occupied the attention of inventive genius. As far back as 1746, at least, the capabilities of the screw as a motive power for ships have been tested by experiments. In 1770 James Watt, who had so much to do with perfecting the steam-engine, suggested the use of screw-propellers. In 1815 Trevethick took out a patent for one. Woodcroft did the same in 1826; but it was not until ten years later that its utility was successfully demonstrated.

In 1836 Captain John Ericsson, a Swede, then residing in London, and Mr. T. P. Smith, of the same place, almost simultaneously had each small boats built for the purpose of testing the screw. Ericsson's boat, named the *Francis B. Ogden*, was 45 feet long and 8 feet beam, and was fitted with two screw-propellers attached to the same shaft. The first experiment made on the Thames was successful beyond all expectation, for he towed the Admiralty barge, with a number of their Lordships on board, from Somerset House to Blackwall and back, at the rate of ten miles an hour. Smith's boat was equally successful, the immediate result being the formation of a joint stock company, called the Screwship Propeller Company, who bought out Mr. Smith's patent and proceeded to build the *Archimedes*, a vessel of 237 tons, and 80 horse-power. Smith's original propeller was a genuine screw, with two whole turns of the thread, made to revolve rapidly under water in the dead-wood of the vessel's run. In the meantime, about 1838, Mr. James Lowe obtained a patent for an important modification of the elongated screw-propeller. This consisted in making use of curved blades, each a portion of a curve, which, if continued, would form a complete screw. The "pitch of the screw" being the whole length along the spindle shaft of one complete turn of the screw, if fully developed, it was found that by reducing the pitch to a segment of the screw and increasing the diameter, the propeller could be reduced to more convenient dimensions.

The success of the *Archimedes* at length induced the Admiralty to make trial of the screw in the Royal Navy. The first *Rattler* was built in 1841, and fitted with a screw-propeller. In 1842 the United States Government made a similar experiment with the *Princeton*, and in the following year the French Government built the screw warship, *Pomone*.[*] In each case the verdict was favourable to the introduction of the screw in preference to the paddle-wheel. The second *Rattler*, of 880 tons and 496 horse-power, was built and fitted with a screw-propeller, and attained a speed of $9\frac{1}{4}$ knots on her trial trip, September 5th, 1851. That settled the question in so far as the Royal Navy was concerned. In the mercantile marine the *Great Britain* was the first ship of large dimensions in which the screw was adopted. For many years there continued to be a strong prejudice against it, though it was destined eventually to entirely supersede the paddle on the ocean.

In order to prevent the screw "racing," which often occurs in heavy weather, to the discomfort of

---

[*] "Our Ocean Railways," p. 75.

passengers and the annoyance of engineers, a system of raising and lowering the propeller has been tried somewhat extensively in the navy and also in the mercantile service, but it has been practically abandoned since the twin screws have come into general use, by which the difficulty alluded to has been largely overcome.

A MYTHICAL WIND-BOAT, FROM AN OLD ENGRAVING (1805).

CHAPTER III.

THE CUNARD LINE AND ITS FOUNDERS.

THIS well-known line takes its name from Samuel Cunard (afterwards Sir Samuel), a native of Halifax, Nova Scotia, who had for some time been conducting the mail service between Halifax, Boston, Newfoundland and Bermuda, and who had long been revolving in his mind the idea of establishing a regular line of ocean mail steamers, but could not find the necessary financial backing in his native country. Proceeding to Britain, Mr. Cunard fortunately fell in with Robert Napier, the famous Clyde ship-builder and engineer, who entered heartily into his proposals and introduced him to George Burns (afterwards Sir George), one of the foremost men in shipping circles at that time, and a man of large means. Through him Mr. Cunard was introduced to David MacIver, of Liverpool, who was of a kindred spirit. The result before long was a partnership of these three with a subscribed capital of £270,000 sterling, and the obtaining of a contract with the British Government for seven years to institute and maintain a steam service from Liverpool to Halifax and Boston, twice a month during eight months of the year and once a month in winter, for an annual subsidy of £60,000. Subsequent stipulations made by the Admiralty were accompanied by an increase of the subsidy to £80,000. At the end of seven years the contract was renewed, but for a weekly service in summer, and twice a month in winter. Saturday then became the regular

"BRITANNIA," FIRST OF THE CUNARD LINE, 1840.

day of sailing from Liverpool, and New York was adopted as one of the American termini. In 1848, when it was found that a weekly service was required, the subsidy was increased to £156,000 per annum. In 1860, to facilitate the despatch of the mails, the boats began to call at Queenstown both going out and returning home, as they still continue to do. In January, 1868, a new mail contract came into operation, under which the Cunard Line received £70,000 a year for a direct weekly service to New York. In the following year Halifax was left out of the programme, although a separate branch line continued to run to Boston as it still does.

The original name of the company was "The British and North American Royal Mail Steam-Packet Company," but it soon took the less cumbrous title of "The Cunard Steamship Company, Limited." The Cunard Line commenced its service from Liverpool to North America on the anniversary of American Independence, the 4th of July, 1840, superseding as mail-carriers the ten-gun sailing brigs of earlier days.*

The first fleet consisted of four side-wheel steamers, each 207 feet long, $34\frac{1}{3}$ feet beam and $22\frac{1}{2}$ feet deep. Their wooden hulls were constructed by four different builders on the Clyde—the *Acadia* by John Wood, the *Britannia* by Robert Duncan & Co.; the *Caledonia* by Charles Wood, and the *Columbia* by Robert Steele. All four were built after the same model, closely resembling that of the *Great Western*. They were all supplied with engines of the side-lever type, by Robert Napier & Sons, 403 horse-power, nominal, with cylinders of $72\frac{1}{2}$ inches diameter and 82 inches

---

* For at least a hundred and fifty years the Post Office Department had maintained a fleet of armed mail "packets." They had stations at Dover, Harwich, Holyhead, Milford, Yarmouth and Falmouth, the last-named being the headquarters of the fleet. During the time of the American war, 1812–15, no fewer than thirty-two sanguinary battles were fought with American privateers by the Falmouth packets, which, in a majority of instances, successfully resisted their assailants.

stroke. They burned about forty-four tons of coal per day, and carried a steam pressure of 9 pounds to the square inch. The *Britannia*, commanded by Captain Woodruff, R.N., sailed on her first westward voyage on July 4th, and after calling at Halifax, reached Boston on the 19th, having made the passage in 14 days, 8 hours, including detention at

THE "NIAGARA," AS A TRANSPORT IN 1855.

Halifax. So great was the enthusiasm in Boston, it is said that Mr. Cunard, who had come out in the *Britannia*, received eighteen hundred invitations to dinner during the first twenty-four hours of his stay in the city! From that time until now the service has been maintained with marvellous regularity, and the line has an unrivalled reputation for safety.

During all these intervening years the ships of the Cunard Line have crossed and recrossed the stormy Atlantic without the loss of a single life. In the early days of the service, the *Unicorn*, formerly of the Glasgow and Liverpool Line, plied between Quebec and Pictou, N.S., in connection with the Atlantic steamers, and is said to have been the first transatlantic steamer to reach Boston, on June 2nd, 1840. The *Unicorn* was commanded by Captain Walter Douglas—a great favourite with his passengers—and the boat was a very fine one indeed.

The second contract, calling for weekly sailings, necessitated a larger fleet of steamers. To meet this demand four new ships were built, and took their places on the line in 1848, namely, the *America*, *Niagara*, *Canada* and *Europa*. Each of these was 251 feet long, of 1,800 tons burthen and 750 horse-power. They had an average speed of $10\frac{1}{2}$ knots an hour. And so, from time to time, as the exigencies of trade and the need for enlarged passenger accommodation demanded, fresh additions were made to the fleet, each succeeding ship surpassing its predecessors in size, equipment and speed. The *Persia*, built in 1856, was the first of the iron boats; the *Scotia*, in 1862, was the last of the paddle-wheel steamers. They were both very fine ships of 3,300 and 3,871 tons, respectively, accounted the best specimens of marine architecture then afloat. The *China*, launched in 1862, was the first Cunard single-screw steamer. She was followed, in 1867, by the *Russia*, the queen of ocean steamers in her day. Passing a number of intervening ships, we come, in 1881, to the *Servia*, the first of the line built of steel—a magnificent vessel, 515 feet long, 7,392 tons, 9,900 horse-power, and attaining a speed of 16.7 knots.

In the meantime important changes had been transpiring in the constitution of the Cunard Company and its environment. The original shareholders had been by degrees bought out by the founders, so that the whole concern was vested in the three families of Cunard, Burns, and MacIver. Sir Samuel attended to the business in London, Mr. Burns in Glasgow, and Mr. MacIver in Liverpool, and never was any business better managed than by these men and their successors. In 1878 it was deemed expedient to consolidate the interests of the partners by the formation of a joint stock company with a capital of £2,000,000 sterling. The three families interested in the concern took up £1,200,000 in paid-up shares No shares, however, were offered to the public until 1880, when a prospectus was issued, setting forth the necessity for additional steamships of the most improved type, involving a large outlay of money. The shares were readily bought up and measures were taken to increase the efficiency of the fleet, which had become at length imperative owing to the keen competition of rival lines. This was inevitable.

The manifest success of the Cunard Company could not long continue without exciting competition, and this followed in due course from a variety of quarters; nor was it to be expected that they should easily hold the supremacy of the sea against all new-comers. They had, in fact, to contend vigorously for their laurels, and at successive intervals had to retire into the second rank, but their determination to regain and hold, at whatever cost, the championship has been well illustrated in the newer ships of the line. The *Umbria* and *Etruria*, steel ships launched in 1884, having cost nearly two millions of dollars each, were a decided advance upon any

THE "SCOTIA," LAST OF CUNARD PADDLE-STEAMSHIPS, 1862.

steamers then afloat. They are 500 feet long, 57 feet 3 inches wide, and 40 feet in depth; they are of 8,127 tons, 14,500 horse-power and are equal to a speed of $19\frac{1}{2}$ knots an hour. They have ample accommodation for 550 first-class passengers and 800 steerage. Each of them has made the run from Queenstown to New York (2,782 knots) in less than six days. In nine consecutive voyages the *Etruria* (in 1885) maintained an average speed of 18 knots.

Her fastest voyage, however, from Queenstown to New York, was made in August, 1897, when she was thirteen years old—namely, 5 days, 21 hours and 10 minutes actual time, the average speed during the voyage being about 20 knots.

It helps one to understand the enormous cost of such vessels when it is stated that the single screw-propeller weighs about thirty-nine tons and costs $25,000! Splendid as was the record of these crack Cunarders, they were surpassed by ships of the White Star and Inman Lines. Something had to be done. An order was given to the Fairfield Ship-building and Engineering Company on the Clyde to build two steel twin-screw express steamships that should surpass all previous efforts. The result was the *Campania* and *Lucania*, launched at Govan in September, 1892, and February, 1893, respectively. These sister ships are splendid specimens of marine architecture. They are each 620 feet long, 65¼ feet beam, and 43 feet in depth. Their gross tonnage is 12,950 tons; their twin screws are driven by triple expansion engines of 30,000 indicated horse-power. Each engine has five cylinders and three cranks. The low-pressure cylinders have the enormous diameter of 8 feet 2 inches; the two high-pressure cylinders are 37 inches in diameter, and the intermediate are 79 inches, with a stroke of 5 feet 9 inches. They are arranged tandem fashion, with a high-pressure cylinder over a low-pressure cylinder, one at each end, and the intermediate in the centre. At eighty revolutions (their normal speed) this enormous weight is moved about

THE "CAMPANIA," AT LIVERPOOL LANDING-STAGE.

2,000 feet per minute. The crank shaft is twenty-six inches in diameter, and each of the three interchangeable parts weighs twenty-seven tons. The propeller shaft is twenty-four inches in diameter, fitted in lengths of twenty-four feet, each length having two bearings. The bossing out of the stern, as in the *Teutonic* and *Majestic*, permits the screws to work without any exterior overhanging bracket, as in other screw steamers. The central boss of the propeller is made of steel; the three blades, weighing eight tons each, are of manganese bronze. A new feature in the machinery is what is called an "emergency governor," which, in case of the shaft breaking, or the screw racing from any other cause beyond a certain speed, is designed to act automatically on the reversing gear and stop the engines. These gigantic engines are started and reversed by steam. Their height from the base to the top of the cylinders is no less than forty-seven feet. There are twelve large boilers, with four furnaces at each end, and made to stand a pressure of 165 lbs. to the square inch. The two funnels are each twenty feet in diameter, and rise to a height of 130 feet above the floor of the ship. The rudder is one large plate of steel, 22 x 11½ feet in area and 1½ inches thick. With the steering gear it weighs forty-five tons! On her maiden voyage from New York to Liverpool the *Campania* eclipsed all previous records, making the run to Queenstown, by the long route (2,896 knots), in 5 days, 17 hours, 27 minutes. Her fastest eastern passage has been 5 days, 9 hours, 18 minutes, and westward, 5 days, 9 hours, 6 minutes. She has run 548 knots in twenty-four hours, and maintained an average speed of 21.82 knots an hour throughout an entire voyage.

Wonderful as the performances of the *Campania* have been, they are surpassed by her sister ship. The *Lucania* made the western voyage, from Queenstown to New York, arriving October 27th, 1894, in 5 days, 7 hours, 23 minutes, the fastest voyage between these points yet made. Her daily runs on that occasion were, 529, 534, 533, 549, 544, 90—total knots, 2,779. Her fastest eastward voyage (up to July, 1897) has been 5 days, 8 hours, 38 minutes; her best average speed throughout a voyage was 22.1 knots an hour, and her highest day's running is 560 knots. The arrival and departure of these steamers at the Liverpool landing-stage has come to be anticipated with almost as much exactitude as that of our best regulated railways. The mails which they carry from New York on Saturday morning are usually delivered in Liverpool on the following Friday afternoon, and letters from London are delivered in Montreal in seven days. By arrangement with the Admiralty, and in consideration of an annual subvention of £19,000, the *Lucania* and *Campania* are held at the disposal of the Government whenever their services may be required as armed cruisers. Other ships of this line are also at the disposal of the Admiralty without any specified subsidy.

Changes and improvements of very great importance to the travelling community have taken place within the last few years, not only in regard to the ocean steamships, but also in regard to facilities for embarkation and landing, and this very largely owing to the lively competition of Southampton and the inducements which it has to offer as a shipping port. The dredging of the bar at the mouth of the Mersey, so as to admit of sea-going vessels entering the port at any state of the tide, is not the least important of the changes referred to. Until quite recently ocean steamers had frequently to come to anchor six or eight miles from the mouth of the river, and wait outside for hours till the tide would rise. That obstruction has been removed, and now the largest steamers can cross the bar at almost any state of the tide. But that is not all. The tedious and discomfortable method of being conveyed from ship to shore in a "tender" has also been done away with. The wonder is that it was submitted to so long. The ocean steamship on her arrival at Liverpool is now brought alongside the landing-stage, and instead of being obliged to drive in a cab or omnibus across the city a mile or more to the railway station for London or elsewhere, the railway and the station have come down to the water's edge, and you pass at once from the ship to the railway train, and immediately proceed on your journey. Passengers for New York may leave Euston Station, London, at noon by a special train of the London and North Western Railway, and find themselves on the landing-stage at Liverpool at 4.15 p.m., the run of over two hundred miles being made, perhaps, without a stoppage—

looking for their luggage, as Englishmen are accustomed to do, and astonished to learn that, by some occult system of handling, and, most strange of all, without a "tip," it is already on board the ship!

Each of these ships is designed to carry six hundred first-class and over one thousand second and third-class passengers. The accommodation provided for them are of the most elaborate description. No expense has been spared in the internal fittings of the ships. Everything that science and skill and refined taste could suggest has been brought into requisition. A more facile pen than ours describes the public rooms, as we call them, as follows, in terms by no means too appreciative: "The dining saloon is a vast, lofty apartment near the middle of the ship, one hundred feet long, sixty-two feet broad, and ten feet high, capable of seating at dinner 430 passengers in their revolving armchairs. The decorations are highly artistic. The ceiling is panelled in white and gold, the sides in Spanish mahogany, and the upholstering is in a dark, rich red, figured frieze velvet, with curtains to match. There are nooks and corners where small parties may dine in complete seclusion. The forty side-lights are of unusual size. Fresh air is admitted by patent ventilators in the roughest weather. For lighting, as well as ventilation, there is an opening in the ceiling in the centre of the room, 24 x 16 feet, surmounted by a dome of stained glass reaching a height of thirty-three feet above the floor. The drawing-room is a splendid apartment, 60 x 30 feet. The walls are ornamented with satin wood, richly carved. The furniture is upholstered in rich velvets and brocades. In the cosy fireplace there is a brass grate and a hearth laid with Persian tiles. The ceiling is in pine, decorated in light tones, old ivory prevailing, with not too much gilding. A Grand piano and an American organ are also provided. The library, 29 x 24 feet, is very ornate. It is suitably furnished with writing tables and writing materials, and a handsome book-case filled with a choice selection of books The smoking-room, 40 x 32 feet, is decorated in the Scottish baronial style. The whole tone of the room is suggestive of *otium cum dignitate.* The ordinary staterooms are lofty and well ventilated, with cunning devices for the saving of room and making things look pleasant and comfortable. Then there are suites of rooms elaborately furnished with tables and bedsteads and bath-rooms, and every conceivable luxury of that sort, for those who are able and willing to pay for them." The accommodation for second-class passengers is in keeping with that for the first. These, too, have their elegant dining-room, and drawing-room, and smoking-room. Even the third-class can rejoice with their neighbours in "the comforts of smoke."

One of these ships, when carrying her full complement of passengers, will start on her voyage provisioned somewhat on this scale: 20,000 lbs. of fresh beef, 1,000 lbs. of corned beef, 10,000 lbs. of mutton, 1,400 lbs. of lamb, 500 lbs. of veal, 500 lbs. of pork, 3,500 lbs. of fresh fish, 1,000 fowls—400 chickens, 250 ducks and geese, 100 turkeys, 30 tons of potatoes, 30 hampers of vegetables, 18,000 eggs, 6,000 lbs. of ham, 3,000 lbs. of butter, etc., etc.; 13,650 bottles of ale and porter, 6,650 bottles of mineral waters, 1,600 bottles of wines and spirits, are frequently consumed on a single voyage.

The various vessels of the Cunard fleet between them carry on an average 110,000 passengers per annum, besides 600,000 tons of merchandise and 50,000 carcases of dead meat in refrigerators, over a distance of one million miles annually. The *Campania* and *Lucania*, owing to the large space occupied by their machinery, only carry about 1,600 tons of freight each.

The order and discipline on board a Cunard liner is that of a man-of-war. The vessels have been built under a special survey, and combine in their construction the best known appliances, in cases of fire, collision, or any other marine contingency, for the safety of the ship and its living freight. The watertight bulkheads are sixteen in number, and will enable the ship to float with any two or even three of the compartments filled with water. The life-boat equipment and service is ample and thoroughly organized. In short, everything is made subservient to safety.

Some idea of the cost of running vessels of this size and speed may be formed when it is stated that the daily average consumption of coal is nearly four hundred tons, but when urged to utmost speed it would be nearer five hundred tons. The crew, all told, number about 424, of whom 195 are required to attend to the engines and boilers alone. In the sailing department, from the captain to the lamplighter, about sixty-five; in the steward's department, including 8 stewardesses, about 120, and in the cook's department, about 45. These 424 persons must be paid and fed at a cost of from $12,000 to $15,000 a month. Each of the ships must have cost over $3,000,000, the interest upon which, at four per cent., is $120,000 per annum; add the enormous cost of provisioning the ship for perhaps six hundred cabin passengers, who, for the most part, expect to fare more sumptuously every day they are on board than they do at home; and one thousand intermediate and steerage passengers, who must live like fighting-cocks; then estimate, if you can, the cost of insurances, agencies, advertising, port charges, pilotage; write off a reasonable percentage for wear and tear; these put together represent an amount so formidable as to leave a very slender margin for profits. At the last annual meeting of the shareholders a dividend of $2\frac{1}{2}$ per cent. for the year 1897 was declared, which was considered a good showing.

Since 1840 the Cunard Company have employed no less than fifty-six first-class passenger steamships in the Atlantic service alone. The entire fleet at present consists of thirty-three ships, with a total tonnage of 124,124, and 153,732 horse-power, and maintains regular communication from Liverpool to New York, Boston, France and almost every country in the Mediterranean. Excepting some of the ships acquired by purchase, all the others were built to order on the Clyde. In all these fifty-eight years the Cunard Company has only lost three ships. Through the mistake of her pilot, the *Columbia*, one of the first Atlantic fleet, ran ashore during a fog near Cape Sable, N.S., in July, 1843, and became a total wreck, but her mails and passengers were safely landed. In 1872 the *Tripoli*, of the Mediterranean Line, was wrecked on the Tuskar Rocks in St. George's Channel, half-way between Cork and Dublin, but no lives were lost. In 1886 the company met with its severest loss by the sinking of the magnificent steamship *Oregon*, recently purchased from the Guion Company. Early in the morning of the 4th of March she was run into by an unknown sailing vessel when about fifty miles from New York, and such were the injuries she sustained she gradually filled with water and went to the bottom, not, however, before the whole ship's company, numbering 995 souls were safely transferred to the *Fulda* of the North German Lloyd Line, which fortunately came up to the scene of the disaster in the nick of time. Her bulkheads should have saved her from going under, and would have done so, but for some unexplained obstruction to the closing of a water-tight door. As it was, the bulkheads kept her afloat long enough to save the lives of all on board.

Among the famous captains in the forties were C. H. E. Judkins, James Stone, William Harrison, Ed. G. Lott, Theodore Cook, Captain Moodie, and James (afterwards Sir James) Anderson who commanded the *Great Eastern* on some of her cable-laying expeditions. Captain Harrison was the first commander of the *Great Eastern*, and was drowned in the Solent when going ashore from his ship in a dingy. Captain Judkins was born at Chester in 1811; he entered the Cunard service in 1840 as chief officer of the SS. *Acadia:* was appointed commander of the *Britannia* that same year, and was successively master of the *Hibernia, Canada, Persia* and *Scotia*. He lived to be Commodore of the fleet and retired from the sea in 1871, after having made more than five hundred voyages across the Atlantic without any serious accident, and being able to say that the Cunard Company at that date had lost neither a life nor a letter. Captain Judkins died in 1876. He was a typical British sailor. He could be exceedingly gracious, and when the mood struck him he could be gruff. I remember making a voyage with him on the *Hibernia* in 1843, on which occasion he ran across from Halifax to Liverpool under a cloud of canvas, with studding sails set low and aloft most of the time, a dense fog all the way, but he picked up his pilot off Cape Clear, just where he expected to find him, and went snoring up the Channel, growling like a bear at the captain of a Dublin steamer who would not get out of his way, and whom in his wrath he threatened to send to "Davie Jones' locker." The voyage was made in nine days and a half, I think, which was accounted a marvellous run in those days. Captain Lott was one of the most genial of men and very popular. He, too, was banqueted on the completion of his five hundredth trip. It has been said of him that his good nature was occasionally ruffled when liberties, unconsciously or otherwise, were taken with his name; as, for example, when a worthy minister officiating on board took for his text, "Remember Lot's wife"; and again, when a rough sailor complained in his hearing that his pork was "as salt as Lot's wife."

Sailors, as a rule, are not given to talk shop, and are quick to resent idle talk in others. The story is told of Captain Theodore Cook that one day when taking his noon observation, a cloud interrupted his vision. Just then a passenger coming along said with a patronizing air, "Captain Cook, I'm afraid that cloud prevented you from making your observation." "Yes, sir," replied the potentate of the sea, "but it did not prevent you making yours."*

At the time of the "*Trent* Difficulty," as it was called, in 1861, the *Australasian* and the *Persia* of the Cunard Line were chartered by the British Government to bring out troops to Canada. On the 4th of December orders were received to prepare the *Australasian* with all speed for this service; her fittings were completed on the 10th, she took in her coal on the 11th, and sailed on the 13th with the 60th Rifles. On the 5th of the same month similar orders were received for the *Persia*, which sailed on the 16th with 1,180 troops, consisting of 1st Battalion of the 16th Regiment and a detachment of sappers. Captain Cook, of the *Australasian*, having encountered much ice in the entrance to the Gulf of St. Lawrence, had to turn back, and took his

* Sir John Burns in *Good Words* for 1887, p. 261.

ship to Halifax and thence to St. John, New Brunswick, where he landed his contingent. Judkins, on the other hand, brought the *Persia* right up to Bic and landed his men, but, the ice threatening to keep him there, he quickly bolted for the open sea, leaving his boats behind him!

Of the more recent commanders, Captain W. H. P. Haines, late of the *Campania* and Commodore of the Cunard fleet, may be instanced as a good specimen. A born sailor he may be called, inasmuch as he is a native of Plymouth, whose father and grandfather before him followed the sea, and who himself has been sailing for nearly fifty years and counts 592 voyages across the Atlantic. Captain Haines has always been as noted for caution as for skill. It is said of him that "whatever temptation there might be to make a fast passage, he would never neglect to take soundings, nor rely on any patent apparatus, without repeatedly fortifying its results by stopping his ship to get up and down casts with the ordinary lead."

To guard against the risks of collision with other vessels, the Cunard steamers follow prescribed routes laid out for them, by which the ships, both outward and homeward bound, are kept at a respectable distance. In estimating the runs of the Atlantic liners from Liverpool to New York and return, Daunt's Rock, off Queenstown, and the Sandy Hook lightship, twenty-six knots from New York, are regarded as the points of departure and arrival; but as Daunt's Rock is about 244 knots from Liverpool, it follows that, to complete the voyage, a full half day's run

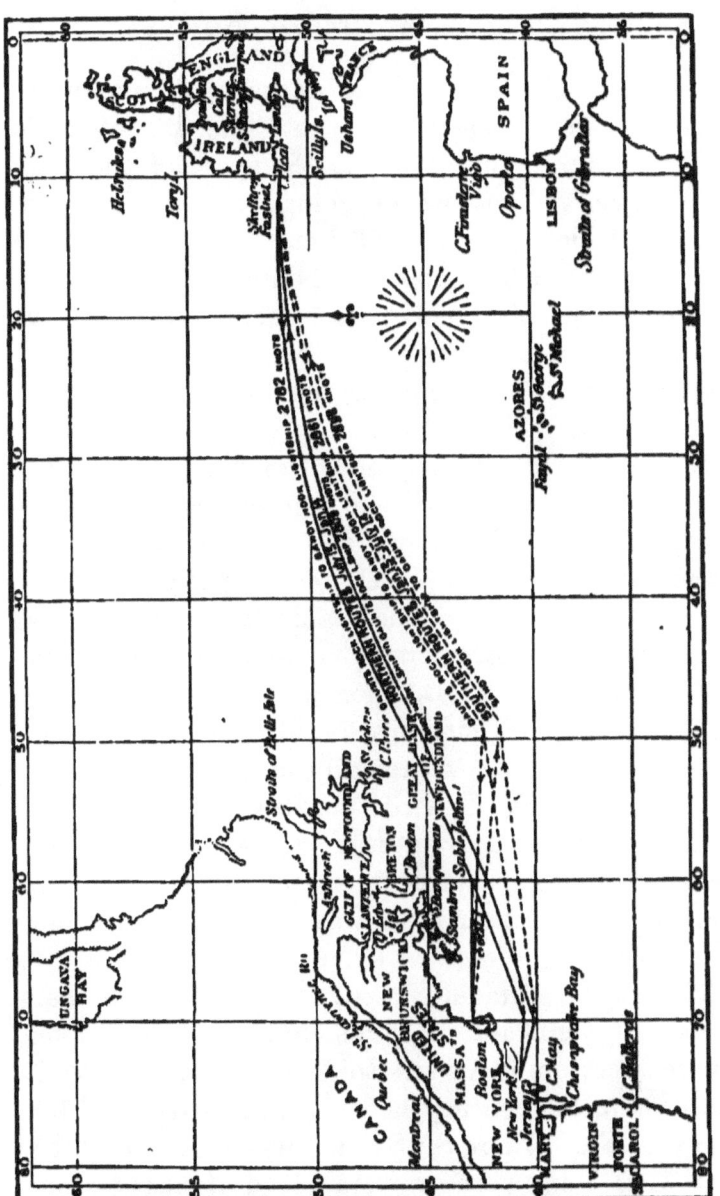

CUNARD TRACK CHART.

must be added to the record as usually announced. It is also to be remembered that the day at sea is longer or shorter according to the speed of the ship. On a twenty-knot vessel going east the average length of day is about 23 hours and 10 minutes; going westward it is about 24 hours and 50 minutes. The difference of time between Greenwich and New York is about five hours.

The "express steamers," as the fast ships are now called, of the Cunard Line at present are the *Campania, Lucania, Etruria* and *Umbria*. These four constitute the weekly mail service, sailing every Saturday from Liverpool and New York. The *Aurania, Servia* and other vessels perform a fortnightly service from the same ports, sailing on Tuesdays. Five steamers are employed in maintaining a weekly service between Liverpool and Boston, and about a dozen more are required for the service between Liverpool, France and the Mediterranean.

The story of the Cunard Company would be incomplete without at least a brief reference to its three founders, Messrs. Cunard, Burns and MacIver, and Mr. Napier, the engineer, who was so closely identified with them.

The late Sir Samuel Cunard was a son of Abraham Cunard, a merchant in Philadelphia, and a Quaker, whose ancestors had come to America from Wales in the seventeenth century, and who removed to Halifax, Nova Scotia. There Sir Samuel was born, November 21st, 1787. His parents were not in affluent circumstances; indeed he has been heard to tell that, when

THE FOUNDERS OF THE CUNARD LINE.

a boy, he often went about the streets with a basket on his arm selling herbs that were grown in his mother's garden, to earn "an honest penny." In course of time, however, he became a prosperous merchant and the owner of whaling-ships that sailed from Halifax to the Pacific Ocean. How he came to identify himself with the Atlantic mail service has already been mentioned, and little else remains to be said about him. He was small of stature, but a man of rare intelligence; a keen observer of men and things, and who had the faculty, largely developed, of influencing other men. In private life he was one of the most gentle and lovable of men. He married, in 1815, a daughter of Mr. W. Duffus, of Halifax, by whom he had nine children. On March 9th, 1859, Her Majesty, on the recommendation of Lord Palmerston, made him a Baronet, in recognition of his services to the realm and to other countries in promoting the means of intercommunication. He was elected a Fellow of the Royal Geographical Society in 1846. He died in London, April 28th, 1865, leaving, it is said, a fortune of £350,000. His title and his interest in the business were inherited by his eldest son, Sir Edward Cunard, at whose decease, in 1869, the reins of administration fell into the hands of his brother William, who married a daughter of the late celebrated Judge Haliburton, of Nova Scotia, and who now represents the company in London.

Sir George Burns was, in many respects, a remarkable man. He was born in the Holy Land, a name popularly given to a "land" of houses in Glasgow, in which five ministers resided, one of them being his father, the Rev. John Burns, D.D., of the old Barony parish, who ministered in that place for seventy-two years, and who died at the patriarchal age of ninety-six. George was born in 1795. He commenced business in Glasgow with his brother James, under the firm of G. & J. Burns & Co., a name that has ever since been famous in shipping circles. They began steam navigation to Liverpool and Belfast over seventy years since, and gradually built up a large and lucrative business. Many years ago Mr. Burns retired and took up his residence at Wemyss Bay, on the estuary of the Clyde, where he spent the evening of his days, and was frequently seen sitting among his rhododendrons and laurels, watching his steamers as they coursed up and down the Firth. He was created a Baronet in his old age, May 24th, 1889. He died on the 2nd of June in the following year, being succeeded by his son, Sir John Burns, of Castle Wemyss, who is chairman of the Board of Directors of the Cunard Steamship Company. Sir John's elevation to the peerage, at the time of Her Majesty's Diamond Jubilee, when he assumed the name of Lord Inverclyde, was regarded as a well-merited honour by his countrymen, and in shipping circles generally.

Although he was a son of the "Father of the Church of Scotland," Sir George early in life contracted a liking for the liturgical service of the Church of England, and eventually became an

Episcopalian. "Sir George Burns, Bart.: His Times and Friends, by Edwin Hodder; Hodder and Stoughton, London," is the title of an admirable biography in which is to be found a fine portraiture of a man "diligent in business, fervent in spirit, serving the Lord." As a business man he is described as "honourable in the minutest particulars, accurate in all his dealings, faithful to every trust, tenacious of every promise, disdaining to take the least advantage of the weakness or incapacity of any man." There is also much information in this volume, bearing on the history of the Cunard Line, that is valuable and interesting, and of which we have freely availed ourselves in compiling these pages.

David MacIver, a Scotchman, as his name implies, had lived a good many years in Liverpool before his connection with the Cunard Company, and had acquired a great deal of valuable experience in shipping affairs. His first intercourse with Burns was somewhat singular in the light of their future alliance. It was as the agent of an opposition line of steamers, plying between Liverpool and Glasgow, that their friendship began. A Manchester firm had started an opposition line, but they were no match for G. and J. Burns, who eventually bought them out, and secured a monopoly of the trade, except the small steamer *Enterprise*, for which David MacIver was agent, and which the same firm cleverly bought also. Not to be outdone, MacIver succeeded in organizing the "New City of Glasgow Steam-Packet Company," of which he became the Liverpool agent. Determined, if possible, to drive his rivals from the seas, it is said that he used to sail in the vessels himself, urging his officers to increased speed. But it was of no use; the new company were soon glad to accept offers for amalgamation, and from that time MacIver and Burns became fast friends. Mr. MacIver had first-rate executive ability, and as most of the working details devolved upon him, he had a controlling influence in the Cunard Line while he lived. The well-known firm of D. & C. MacIver were the managers of the line at Liverpool, from its formation until the year 1883, when they resigned, a Board of Directors assuming the entire control of affairs. David MacIver, however, had died in 1845, when the Liverpool agency fell into the hands of his brother and partner, Charles, whose able supervision continued for thirty-five years.

Robert Napier was born at Dumbarton in 1791. After serving his apprenticeship as millwright and smith, he went to Edinburgh, where he wrought at his trade for some time, earning ten shillings a week. Inspired by the old Scotch motto, "He that tholes overcomes," he stuck to it. Later, he entered the service of Robert Stephenson, the celebrated engineer, and made his mark as a mechanical genius. At twenty-four years of age he commenced business on his own account, in Glasgow, where he gradually built up the large engineering and ship-building business subsequently carried on under the name of Robert Napier & Sons. The "Lancefield Works" and his Govan shipyards attained world-wide celebrity. He constructed the machinery for the SS. *British Queen*, and for the first four Cunard steamers, and for many others in later years. He also received large orders for warships and transports from the British Admiralty and from foreign governments. He built several large ironclads for the Royal Navy. He made the engines for the great three-decker, *Duke of Wellington*—all but the last of the "wooden

ROBERT NAPIER AND MRS. NAPIER.

walls." He built and engined the famous Cunarders *Persia* and *Scotia*.

Mr. Napier erected a princely mansion on the Gareloch, named Shandon House, where his declining years were spent in retirement, but in the exercise of unbounded hospitality, as the writer can testify from his personal experience. Shandon House came to be like a museum containing a rare collection of pictures and antiquities from almost all parts of the world. Among his curios none was more highly prized than his mother's spinning-wheel, and the painting that he valued the most was the portrait of his wife plying the same old-fashioned spinning-wheel, with which she had been familiar from girlhood. Does it not seem like the "irony of fate," and a melancholy commentary on the transitory nature of everything mundane, that this marvellous accumulation of articles of *virtu* was, shortly after Mr. Napier's death, sold by public auction to the highest bidder, and that his palatial residence passed into the hands of a hydropathic company?

Having said so much about the Cunard Line, there is no need to dwell at similar length upon any of the other transatlantic lines of steamers. The history of the Cunard Line is the history of Atlantic steam navigation. It commenced at a time when steam power had only been used as an auxiliary to sails, but when that order of affairs was soon to be reversed. The intervening years have witnessed the transition from wooden ships to iron, and from iron to steel; from the paddle-wheel to the single screw-propeller, and then to the twin-screw; from the simple side-lever engines to the compound, and from the compound to the triple and quadruple expansion engines of the present time. These successive changes, common to all the other important lines of ocean steamers, have resulted in greatly increased speed with economy of fuel. But no one at all conversant with the subject supposes that the limit in either of these directions has been reached. Her Majesty's torpedo boats can easily reel off their thirty knots an hour; why not an express steamer?

The competition for the supremacy of the sea in these latitudes has been both keen and costly, but greatly to the benefit of the travelling community; and it has all along been conducted in an excellent spirit. Circumstances have frequently arisen when it might have been easy to take advantage of a rival, but when it resulted in acts of chivalry. Sir John Burns has mentioned one instance out of many such that have transpired: On a certain occasion the Cunard steamer *Alps* was seized in New York for an alleged infraction of the Customs laws on the part of some of the crew, and before she could be released, security had to be given to the extent of £30,000 sterling; when, "who should come forward and stand security for the Cunard Company but the great firm of Brown, Shipley & Co., the agents of the Collins Line!" Another case in point is connected with the foundering of the Cunard SS. *Oregon*. When the whole of the passengers and crew, to the number of nearly a thousand, had been taken off the sinking ship, and landed in New York by the North German Lloyd SS. *Fulda*, the question having been asked what compensation was demanded, the courteous reply was speedily received: "Highly gratified at having been instrumental in saving so many lives. No claim!" *

* Fry's "History," p. 240.

The Fairfield Ship-building and Engineering Company is one of the most famous of the many eminent ship-building firms in Britain. The yards at Govan on the Clyde occupy an area of sixty acres of ground, and employ from 6,000 to 7,000 men. The shops are fitted with machinery of the most approved description, in which every requisite of marine architecture has a place, where massive plates of steel and iron are clipped, shaped and pierced with rivet holes as if they were only sheets of wax or paper. Here have been built many of the record-breaking ocean greyhounds, as well as armour-plated cruisers for the Royal Navy. The *Arizona*, the *Alaska* and the *Oregon* were built here, and were accounted marvels in their day. The *Umbria* and *Etruria*, the *Campania* and the *Lucania* have secured for Fairfield a world-wide reputation. Ships for Sir Donald Currie's Castle Line, for the Orient and the Hamburg-American lines, not to speak of the Isle of Man steamers, the swiftest coasting steamers of the day, have been built at Govan. Under the name of Randolph, Elder & Company the firm was founded, or rather reconstructed, by the late Mr. John Elder, a man of consummate ability in his profession, who died in 1869 at the early age of forty-five years.

The compound engine, by which steam is made to do double duty, is one of the most important of recent improvements in marine engineering, being the means of largely increasing the motive power and decreasing the consumption of fuel. The successful application of this system to ocean steam navigation is usually attributed to Mr. John Elder, of the above-named firm, who introduced it in some of the steamers of the Pacific Steam Navigation Company as early as 1856.* But it did not come into general use until some years later. The Admiralty, recognizing the importance claimed for the discovery, resolved to test its value, in 1863, by sending three ships of similar size on a voyage from Plymouth to Madeira, two of them being fitted with the ordinary engines of the day, and the third, the *Constance*, with Elder's compound engine. The result placed the superiority of the compound engine beyond question, and led up to the triple and quadruple expansion engine which has revolutionized the ship-building and shipping interests; hence the enormous cargoes carried by ships of the *Pennsylvania* type, with a moderate consumption of fuel and the lowering of ocean freight rates.

Before taking leave of the Cunard Line, it may not be out of place to mention that an employee of that line has the distinction of having crossed the Atlantic more frequently than any other man. One is apt to think of his own voyages—thirty-five or forty—as a tolerably fair showing, but that is as nothing compared with other landsmen. On one occasion the writer sat next to a fine old French gentleman from Quebec who was then making his hundredth voyage; he was an octogenarian. Some years later a Montreal merchant, nearly a quarter of a century younger, informed me that he had crossed the ocean *one hundred*

---

* The invention is claimed for Canada in Chapter X., under the heading of "New Brunswick."

and eighty times! Taking his years into account, surely he must be entitled to wear the blue ribbon. As to sailors, an English newspaper recently offered a prize of £10 to the man who could prove that he had crossed the Atlantic oftenest. The prize was awarded to Captain Brooks, of Alaska, who had made seven hundred trips. In the meantime, however, it transpired that the distinction was due to another "old salt," whose record far outran that of Captain Brooks, but whose modesty prevented him from applying for the prize. The real champion is George Paynter, well known throughout England and America as "the Old Man of the Sea," who recently completed his *eight hundred and fourth voyage* across the Atlantic. Paynter is the officer in charge of the wines and liquors on board the SS. *Etruria*. He is one of the most remarkable men afloat to-day. He has been forty-eight years at sea, of which forty-five have been spent continuously in the service of the Cunard Company, and in all that time he has never encountered either a shipwreck or a cyclone. He is now seventy-five years old, hale and hearty as ever, and this he attributes to his having given up smoking and drinking thirty-one years ago, not having once indulged in either from that time until now.

CHAPTER IV.

NORTH ATLANTIC STEAMSHIP COMPANIES.

The Collins Line.

THE earliest formidable rival to the Cunard Line was the famous Collins Line, founded in New York in 1848, and which derived its name from Mr. E. K. Collins, its chief promoter, who had previously been largely interested in sailing ships, and more particularly in the splendid line of New York and Liverpool packets, popularly known as the Dramatic Line. The Collins Line started with a fair wind, so to speak. It was launched by a wealthy company, amid an outburst of national applause, and was liberally backed by the Federal Government, with an ill-concealed determination to drive the Cunarders from the seas. But the illusion was destined to be soon dispelled, for, as Charles MacIver put it in writing to Mr. Cunard, "The Collins Line are beginning to find that breaking our windows with sovereigns, though very fine fun, is too costly to keep up." Disasters ensued. In ten years the losses had become stupendous, and the enterprise culminated in a total collapse.

The Line began with a fleet of four magnificent wooden paddle-wheel steamships, the *Atlantic, Arctic, Baltic* and *Pacific*, each 282 feet in length, and of 2,680 tons burthen. They were built by W. H. Brown, of New York, and combined in their construction and machinery the then latest improvements. The passenger accommodation was far superior to that of the Cunard steamers of the period. Each of them cost $700,000, an amount so far exceeding the original estimate that the Government had to make the company an advance. The credit of the country being in a sense at stake, provision was made for a liberal subsidy. $19,250 per annum had been the original sum specified for a service of twenty round voyages, but that was found to be totally inadequate, and the Government eventually agreed to increase the subsidy to $33,000 per voyage, or $858,000 per annum for only twenty-six voyages, which was more than double what had been paid to the Cunard Company for a like service. The Collins Line, however, promised greater speed than their rivals, and that counts for much in popular estimation.

The Line soon came into favour, and its success seemed to be assured. The first voyage was commenced from New York by the *Atlantic*, April 27th, 1849. The *Arctic* followed, making the eastward voyage in 9 days, 13 hours and 30 minutes; and the westward, in 9 days and 13 hours from Liverpool. Thus they had broken all previous records for speed which, added to their luxurious appointments, caused them to be loyally patronized by the Americans. For a time they carried 50 per cent. more passengers from Liverpool to New York than their opponents. The last addition to the fleet was the *Adriatic*, in 1857, by far the finest and fastest vessel afloat at that time. She was built by Steers, at New York: was 355 feet

THE "ATLANTIC," OF THE COLLINS LINE, 1849.

long, and 50 feet broad; her gross tonnage being 3,670. Her machinery, which was constructed at the Novelty Iron-Works, New York, consisted of two oscillating cylinders, each 100 inches in diameter, working up to 3,600 indicated horse-power, with a steam pressure of 20 lbs. to the square inch. Her paddles were 40 feet in diameter, and, at seventeen revolutions per minute, gave her a speed of thirteen knots on a daily consumption of eighty-five to ninety tons of coal.

Owing to financial embarrassments, resulting from losses by shipwreck, the company soon after broke up, and the richly-endowed fast line, that was to drive the Cunarders off the ocean, itself came to grief. The *Adriatic* was laid up after making a few fine voyages, and finally came to an ignominious end as a coal-hulk in West Africa. In September, 1854, the *Arctic* collided with a small steamer, the *Vesta*, off Cape Race, in a dense fog, and sank, with the loss of 323 lives. Captain Luce went down with his ship, but rose again to the surface, was picked up by one of the boats and landed in safety. Among those who were drowned were the wife, the only son, and a daughter of Mr. Collins, and many other prominent Americans. The loss of the *Pacific*, which followed two years later, proved the death-knell of the Collins Line. She sailed from Liverpool on June 26th, 1856, in command of Captain Eldridge, with forty-five passengers and a crew of 141, and was never afterwards heard of. The *Atlantic* and *Baltic* were sold and converted into sailing ships.

Mr. E. K. Collins was a native of Massachusetts, where he was born in 1802. When a youth he went to sea as supercargo. Some years later he joined his father in the general shipping business, and eventually became head of the New York firm, celebrated for its magnificent line of sailing packets. He died in 1878.

"CITY OF PARIS," 1889.
Now (1898) a U.S. armed cruiser and renamed *Harvard*.

THE INMAN AND INTERNATIONAL LINE.

This famous Line took its name from William Inman, a partner in the firm of Richardson Bros., Liverpool, in connection with whom he founded this steamship service in 1850, under the title of the Liverpool, New York and Philadelphia Steamship Company. The line began with only two steamers—the *City of Glasgow* and *City of Manchester*—both screw steamships, built by Messrs. Tod and McGregor, of Glasgow. These boats having proved successful and profitable, and especially popular with emigrants, their shipping port was changed from Philadelphia to New York in 1857. In the meantime a number of high-class steamers had been added to the fleet, each improving upon its predecessor, until the line became famous for speed and comfort. The *City of Brussels*, launched in 1869, was the first on the Atlantic to reduce the voyage to less than eight days. This fine ship came to grief through collision with another vessel off the mouth of the Mersey during a dense fog, January 7th, 1883. The Inman Line met with a number of other heavy losses. The *City of Glasgow*, with 480 persons on board, and the *City of Boston* both disappeared mysteriously in mid-ocean; the *City of Montreal* was burned at sea, but all hands were saved; the *City of Washington* and *City of Philadelphia* were wrecked on the coast of Nova Scotia; the first *City of New York* and the *City of Chicago* became total wrecks on the Irish coast, the one on Daunt's Rock near Queenstown, the other on the Old Head of Kinsale in the same neighbourhood.

The *City of Berlin*, which came out in 1875, proved a great success, but later additions, culminating in the new *City of New York* and *City of Paris*, gained this line for a time undisputed supremacy. These twin-screw ships, built by J. & G. Thomson, of Glasgow, are over 500 feet in length, rated at 10,500 tons, and 18,000 indicated horse-power, and have developed a high rate of speed. The *Paris*, as she is now called, made her maiden trip in May, 1889, in 5 days, 22 hours, 50 minutes. Her fastest westward trip was made in October, 1892, viz., 5 days, 14 hours, 24 minutes—the fastest ever made up to that time. The *New York* for some time held the record for the fastest voyage from Southampton to Sandy Hook, made in September, 1894—6 days, 7 hours, 14 minutes. Both ships have met with mishaps: the *New York*, going east, had one of her engines disabled, but completed the voyage with the other, actually running 382 knots in one day with only one engine at work. The *Paris* had a much more alarming accident. The breaking of one of her main shafts set the engine a-racing, and before it could be stopped a rent was made in the ship's hull, the longitudinal bulkhead separating the engine-rooms was broken and both engine-rooms were flooded. The other bulkheads, however, did their duty and kept her afloat until a passing steamer towed her into Queenstown, where the water was pumped out and she proceeded to Liverpool unassisted. Her escape from destruction was marvellous: as it was, the damage to the ship and machinery was

"CITY OF PARIS" HER TWIN SCREWS.
From "Our Ocean Railways."

"CITY OF PARIS"—DINING-ROOM UNDER THE DOME.

enormous. On another occasion the same ship's rudder became disabled in mid-ocean, but by means of her twin screws she was kept on her course and brought safely to port. Since then she has limped across the Atlantic with one engine, owing to a broken shaft.

The Inman Line was the first to introduce the twin-screw in the Atlantic service. It was also the first to place the comforts and conveniences of steam navigation within the reach of emigrant steerage passengers, and by so doing made a distinct advance in the cause of humanity. In 1856-57 they carried no less than 85,000 emigrants.

The Inman Line passed from its founders in 1875, and became a private limited company, which, in 1886, entered into negotiations with the American International Navigation Company, better known as the Red Star Line. At that time the fleet consisted of the *City of Berlin*, *City of Chester*, *City of Chicago*, *City of Richmond* and *City of Montreal*. The *New York* and *Paris* hoisted the American flag in 1893, but the change consequent on their new registration and their re-christening made no change in the name of the company.

In 1892 the company secured a contract for carrying the United States mails, weekly, from New York to Southampton, in consideration of a subsidy, amounting to about $750,000 a year. Southampton was preferred to Liverpool as being much nearer London and as having exceptionally good harbour facilities. The sea voyage, however, is about 200 miles longer than from New York to Queenstown. In terms of their contract, two magnificent twin-screw steamers have recently been added to the fleet, —the *St. Louis* and *St. Paul*, built on the Delaware by Messrs. Cramp and Sons, of Philadelphia. They are claimed to be the embodiment of the finest American skill and workmanship. Over 6,000 tons of steel were used in the construction of the hull of each ship; their length over all is 554 feet, breadth 63 feet, depth 42 feet; their gross tonnage is 11,000 tons and their engines are of 20,000 horse-power. They are designed to carry 320 first-class, 200 second, and 800 steerage passengers, and the arrangements for each class are unsurpassed. The main saloon is 110 feet long by 50 feet wide, with seats for all her cabin passengers at one sitting. It is handsomely decorated and finished in white mahogany, and is well lighted from the sides and a lofty dome overhead. The drawing-room is in white and gold and luxuriously furnished. The staterooms are roomy, well ventilated and fitted up with every convenience necessary to comfort; there are also suites of rooms, comprising bedroom, bath-room and sitting-room, all elegantly furnished. These ships can carry enough coal, cargo being excluded, to cross the Atlantic and return at their highest speed; and at the ordinary cruiser's speed of 10 to 12 knots, they can steam for 66 days without recoaling a distance of 19,000 knots.

Although these fine ships have already suffered several vexatious accidents, none of them have been attended with serious results. They have not yet

"ST. LOUIS."
- Now (1898) a U.S. armed cruiser.

taken the laurels from the *Campania* and *Lucania*, and are not likely to do so, but they have made very good time on the Atlantic. The *St. Louis* made the voyage from New York to Southampton in August, 1895, in 6 days, 13 hours, 12 minutes. The *St. Paul** made the run from Southampton to Sandy Hook, in August, 1896, in 6 days, 57 minutes. Their estimated speed in ordinary weather is 21 knots an hour.

The entire Inman fleet consists of twenty-two ships—all of a high class. They retained the graceful overhanging bow and ship-shape bowsprit with its belongings to the last, but the new steamers of the American Line conform in this respect to the prevailing fashion of the straight stem, first introduced by the Collins Line as being economical of space and every way handier in port. The use of sails in full-powered steamships has been gradually declining for years, and they will soon be a thing of the past. Heavy masts and yard-arms seriously interfere with the motion of a twenty-knot steamship, and except in the case of a breakdown of machinery are seldom of any use, and that contingency has been reduced to a minimum by the introduction of the twin-screw.

THE RED STAR LINE,

originally owned by a Belgian company, is now incorporated with the American and International Navigation Company, and maintains a weekly service

---

* The *St. Paul, St. Louis, Paris* and *New York* have all been taken over by the United States Government and fitted up as armed cruisers, the names of the last two being changed to *Harvard* and *Yale*.

between New York and Antwerp and a fortnightly line from Philadelphia to Antwerp. The fleet consists of nine steamships of from 3,000 to 7,000 tons each—the largest being the *Friesland*, built by Thomsons, Glasgow, and rated at fifteen knots' speed.

The Anchor Line.

This was the first successful line of steamers running from Glasgow to New York, established by Messrs. Handyside and Henderson, of Glasgow, in 1856, though it was not until 1863 that this branch of their business assumed much importance. Since then the trade has developed rapidly, giving employment to a weekly line of steamers, and in summer twice a week. The ships have large carrying capacity, from 3,000 to 5,000 tons and upwards, with good accommodation for passengers at very moderate rates. Among these are the *Furnessia* and *Belgravia*, of over five thousand tons; the *Devonia, Anchoria, Bolivia* and *Circassia*, upwards of four thousand tons each, not to speak of the *City of Rome*, a host in herself. This is one of the handsomest ships afloat, and of large dimensions, being 546 feet long between perpendiculars, and 600 feet over all; her width is 52 feet 4 inches, and her displacement at 25 feet draft of water, 13,500 tons. She is driven by three sets of inverted tandem engines of 10,000 horse-power; her single screw is 24 feet diameter, and the screw shaft 25 inches. She has ample accommodation for 270 cabin passengers and 1,500 steerage: was built in 1881 for the Inman

Line at Barrow-in-Furness, Lancashire, where all the above-named ships were also built, but as she did not come up to the requisite speed she was left in the builders' hands, and was taken over by the Anchor Line. She is not a slow ship, having made $18\frac{1}{2}$ knots on her trial trip, and has crossed the Atlantic in 6 days, 20 hours, 35 minutes. From whatever cause, outsiders look upon her as a sort of "white elephant," unable to compete successfully with the more thorough-paced ocean greyhounds. The entire Anchor Line fleet consists of some thirty-five steamers. The company has had its own share of losses by shipwreck, and more than its share of lives lost. One of the most appalling marine disasters was the sinking of the *Eutopia* of this line in the Bay of Gibraltar, in 1891, from collision with a man-of-war lying at anchor, resulting in the loss of 526 lives.

THE NATIONAL STEAM-NAVIGATION COMPANY.

Although the National Line has not entered into competition with the "greyhounds," it is deserving of notice. It has been in existence since 1863, and has owned some fine ships, and at least one of high speed —the *America*, built on the Clyde in 1883—a ship of 5,500 tons and 7,350 horse-power. She broke the record in June, 1884, making the run home from New York in 6 days, 14 hours, 18 minutes.* She was soon after sold to the Italian Government for a transport. The ships of this line were among the first to have compound engines, and the first to have refriger-

* Fry's "History," p. 193.

ators for the reception of dead meat, and among the first to carry large shipments of live cattle. Years ago they brought out more emigrants than any other line, but they seem to have gone out of that business now, and the ships are run as freighters to London. Four of the company's ships have been lost —one lies submerged near Sandy Hook, one foundered off Cape Finisterre, one was burned at sea, and the fourth, the *Erin*, disappeared without anything having been heard of her. The present fleet consists of eight ships, ranging from 3,750 to 5,300 tons.

THE GUION LINE.

As when a meteor shoots athwart the skies, emitting a blaze of light, and quickly disappears, so was it with the Guion Line at the zenith of its brief and brilliant career. It began in a modest way in 1866, its promoters being Messrs. Williams and Guion, of New York—with a branch firm in Liverpool—these being the owners of the famous Black Ball Line of ships, built especially for carrying emigrants. They had steamers built for themselves with marvellous rapidity, beginning with the *Manhattan* of 3,000 tons—an iron screw steamer built on the Tyne. In 1872 there was added to the then existing fleet of eight powerful ships, each having accommodation for 1,000 steerage passengers, a pair of larger vessels, the *Montana* and *Dakota*. Neither of them, however, proved to be "record-breakers," and both of them were eventually wrecked on the Welsh coast, near the same place, in 1877 and 1880 respectively. The next additions to the fleet were the celebrated *Arizona* and *Alaska*, that for a time took the shine out of everything else afloat. These marvellous ships were built by John Elder & Co., of Glasgow. The former was over 5,000 tons and the latter nearly 7,000. Their engines, respectively 6,000 and 10,000 horse-power, are said to have been the finest ever constructed up to that time; their speed was then accounted quite phenomenal—seventeen and eighteen knots an hour—reducing the time from Queenstown to New York to 6 days, 21 hours, 40 minutes. That was in 1883. The last ship built for the Guion Line was still larger and faster than these. The *Oregon* was 500 feet long, of 7,375 tons, and 13,300 horse-power. In 1883 she still further reduced the record to 6 days, 10 hours, 10 minutes. Soon after this the company became involved in financial difficulties. "Record-breaking" had not proved to be a paying business. The *Oregon* passed into the hands of the Cunard Company, and went to the bottom of the sea as already stated; the *Alaska* and *Arizona* have lain rusting at their moorings in the Gareloch for years past.

The White Star Line.

The Oceanic Steam Navigation Company, Limited —better known as the White Star Line—commenced in 1869, and now occupies a position in the front rank of the great steamship enterprises of the world. It originated with Mr. Thomas Henry Ismay, of Liverpool, who had previously been manager of the White Star Line of sailing clipper ships in the Australian trade. In 1870 Mr. William Imrie, of the late firm of Imrie, Tomlinson & Co., became associated with Mr. Ismay in the management, when the firm took its present name, Ismay, Imrie & Co. Mr. Ismay retired from the firm in 1891, after forty years of active business life, but is still chairman of the White

"OCEANIC," FIRST OF THE WHITE STAR LINE, 1871.

Star Line. Having the financial support of a number of influential shipping men, plans that had been long maturing took effect in 1869, when negotiations were entered into with Messrs. Harland and Wolff, of Belfast, to build a fleet of steamships which should combine the latest improvements, the best possible accommodation for passengers, with a speed that would assure fast and regular voyages. How well those conditions have been secured all who have travelled by the White Star Line can testify.

The first ship of this line to appear in the Mersey was the *Oceanic*, in February, 1871. It was at once seen by her graceful lines that she was "a clipper." Her machinery was the best known up to that time. A new feature was that the main saloon and passengers' berths were placed as near midships as possible, and separate revolving chairs were introduced in the dining-room (a great boon to passengers); a number of other innovations served to attract the notice of the travelling community, while admirable management on shipboard and ashore inspired confidence in the line.

The original fleet consisted of six ships — the *Oceanic, Baltic, Atlantic, Republic, Celtic* and *Adriatic* — all about the same size, close upon 4,000 tons each. In 1874 and 1875, two remarkable vessels, as then accounted, were added to the fleet—the *Britannic* and *Germanic*—by the same builders, with engines from Maudslay, Son & Field. These boats are 468 feet long, of 5,000 tons and 5,000 horse-power. They easily made sixteen knots an hour, burning only 110 tons of coal per day, and were in every way so satisfactory they became very popular. No higher compliment can be paid them than the statement made in 1894 that "they had now been running regularly for twenty years, giving complete satisfaction to the owners and to the public, having still the same engines and boilers with which they started."*

* The *Germanic* has since been overhauled and has now a set of triple expansion engines, making her a seventeen-knot boat. In July, 1895, she crossed from Queenstown to New York in 6 days, 23 hours, 45 minutes.

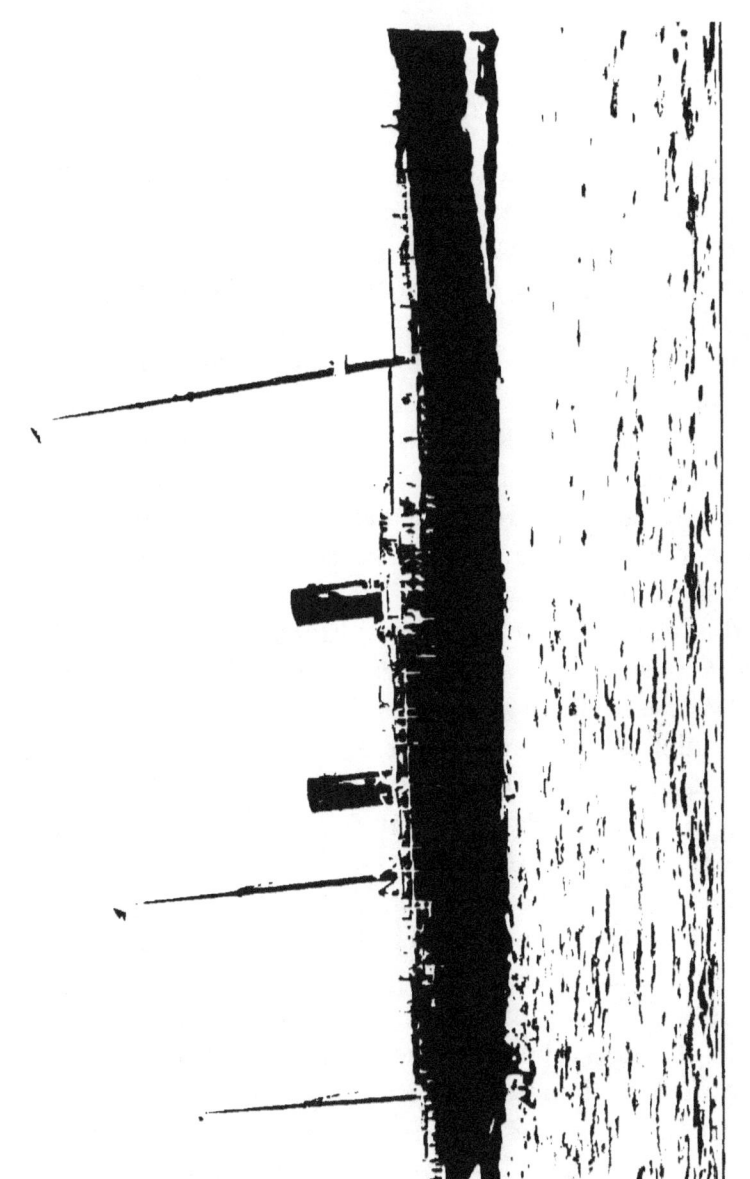

"MAJESTIC," WHITE STAR LINER, LAUNCHED IN 1889.

In those twenty years these two ships carried 100,000 cabin and 260,000 steerage passengers.

In the meantime the new Cunard steamers, *Umbria* and *Etruria*, had outrun the White Star clippers. Again an order was given to Harland & Wolff for a pair of larger, finer and faster boats than they had yet built. The magnificent twin screw steel ships, *Teutonic* and *Majestic*, filled the bill. The *Teutonic* was launched in January, 1889. On the 7th of August she left Liverpool on her maiden voyage to New York, having in the meantime taken part in the naval review at Spithead, where she was inspected and admired by the German Emperor and H. R. H. the Prince of Wales. She crossed from Queenstown to Sandy Hook in 6 days, 14 hours, 20 minutes, then the quickest maiden passage on record. The *Majestic* was launched in June, 1889, and made her first voyage to New York in April following, lowering the record to 6 days, 10 hours, 30 minutes.

These fine ships are each 582 feet in length over all, 57 feet 8 inches in width, and 39 feet moulded depth. Their gross tonnage is 10,000 tons, all to a fraction. They are twin-screw ships, each having two sets of triple cylinders, 43 inches, 68 inches, and 110 inches diameter, respectively, together working up to 18,000 horse-power. The screw-propellers are 19 feet 6 inches diameter, and so fitted that they overlap 5 feet 6 inches, the starboard propeller being six feet astern of the other. They have each twelve double-ended and four single-ended boilers, containing in all seventy-six furnaces. The steam pressure is

180 pounds to the square inch. The piston stroke is five feet, and the average revolutions seventy-eight per minute. About four thousand tons of coal are consumed on the round voyage. Not only do these ships combine in their construction and equipment all that is best in modern improvements, but some of the most valuable of these improvements originated with their builders, and have been largely imitated by others.

The whole service, food and attendance included, is unexceptionable. There is ample accommodation for about 300 saloon, 170 intermediate and 1,000 steerage passengers. As to speed, they "must have swift steeds that follow" them. The *Teutonic* has made the western voyage in 5 days, 16 hours, 31 minutes. The *Majestic* has done it in 5 days, 17 hours, 56 minutes. In ordinary circumstances the passenger who embarks at Queenstown may safely calculate that six days will land him in New York by either of these ships. They are not quite so fast as the *Lucania*, but to gain the difference, say, of ten hours in crossing the Atlantic, the Cunarder requires an enormous increase of driving power—no less than 12,000 horse-power over and above that of the other. The *Teutonic* and *Majestic* are under contract with the British Government to be used as armed cruisers whenever their services may be required, the company receiving an annual sum of £14,659 10s. as a retainer.* Each of these steamers has accommodation for one thousand cavalry and their horses, or for

---

* Fry's "History," p. 180.

2,000 infantry. They could easily reach Halifax from Queenstown in five days, Cape Town in twelve and a half, and Bombay, *via* the canal, in fourteen days from Portsmouth. They could even steam to Bombay, *via* the Cape, 10,733 knots, in twenty-three days without stopping to coal.

The White Star fleet at present consists of nineteen ocean steamers, ranging in size from 3,807 to 10,000 tons and upwards. Five of these steamers are employed in the Atlantic weekly mail service, three keep up a monthly line to New Zealand, four ply monthly from San Francisco to Japan and China, the remainder are cargo boats of large carrying capacity, A number of vessels built for this company have been sold to other lines and are still running. The *Oceanic*, pioneer ship of the line, after a few years in the Atlantic service, was transferred to the company's trans-Pacific line. On her sixty-second voyage in October, 1889, she crossed from Yokohama to San Francisco in 13 days, 14 hours, 4 minutes, the fastest voyage then on record across the Pacific. Having completed twenty-five years of successful work she was sold and broken up in 1896. But the name is to be perpetuated by the magnificent new steamer now building at Belfast, which in point of size and speed is designed to surpass any vessel at present afloat. The new *Oceanic* is longer than the *Great Eastern*.

Only two ships of this line have been lost. The *Atlantic* was wrecked on the coast of Nova Scotia, April 1st, 1873. She had left the Mersey on March 20th, with 32 saloon, 615 steerage passengers, and a crew of 143—790 in all—of whom about 560 perished, including all the women and children. What made the disaster even more deplorable, it was not satisfactorily accounted for. The morning was dark and boisterous, but not particularly foggy. Captain Williams had mistaken his reckoning, and was rushing his ship incautiously too near the land.* The *Naronic* was a fine new cargo ship of 6,594 tons. She left Liverpool, February 11th, 1893, bound for New York; but she never arrived there. Two of her boats were picked up on March 4th, but no clue was ever found to the mysterious disappearance of the ship.

Thomas H. Ismay, recently retired from business, has all along been recognized as the manager-in-chief and moving spirit of the White Star Line, and a man of exceptional gifts and graces. Conspicuous alike by his enterprise and culture, Mr. Ismay has given proof of true greatness in declining honours that were easily within his reach. He might have been chairman of the London and North Western Railway

---

* A missionary of the Church of England, who ministered to a few poor fishermen at Terence Bay, at the imminent risk of his life put off to the wreck in a small boat and succeeded in saving the life of the first officer of the ship after all hope of further rescue had been abandoned, and when even the hardy fishermen forbade the rash attempt. Mr. Ancient had formerly been attached to the British navy, and during this heart-rending scene acted the part of a hero in his efforts to save life and to relieve the sufferings of the survivors. Captain Williams was severely censured, and had his certificate suspended for two years.

Company—the greatest railway company in the world—but he would not. Several times he might have been returned to Parliament, but he declined. His name was confidently mentioned in connection with the Diamond Jubilee honours. Sir Thomas Ismay would have sounded well, but he begged to be excused, choosing to remain plain Thomas Ismay, of Liverpool, where his beneficent character is known and appreciated at its full value. The same may be said of the genial ex-captain of the *Majestic*, and commodore of the fleet, Captain Parsell, in whose personality were combined the culture of a gentleman and all the qualifications of a good sailor. Captain Cameron, of the *Teutonic*, has been in the service of the White Star Company nearly thirty years, having commenced his career in the sailing ships. He is one of the most popular commanders on the route.

Messrs. Harland and Wolff, of Belfast, the builders of all the steamers of the White Star Line, are one of the largest ship-building firms in the world. They employ between seven thousand and eight thousand men in their establishment. Sir Edward J. Harland, late head of the firm, was a Yorkshireman by birth. He served an apprenticeship to engineering at Newcastle, and studied the art of ship-building in the drawing office of Messrs. J. & G. Thomson, Glasgow. He was a man of noble presence, fine ability, and great enterprise. He had been Chairman of the Harbour Board, Mayor of Belfast, High Sheriff of

County Down, a Justice of the Peace, and a member of Parliament. He was made a Baronet by the Queen, in 1885, on the occasion of the visit of the Prince and Princess of Wales to Belfast. Sir Edward died at his home, Glenfarne Hall, County Leitrim, December 23rd, 1895, aged sixty-four years.

The rates of passage by the Cunard, the White Star and the American Line are nearly identical, and, all things considered, they are not unreasonable. They are cheaper than the fares by the sailing packets of sixty years ago. The ordinary rates for first-class passengers, in summer, vary from $75 to $150, according to the location of the stateroom, and the number of berths in it; from $40 to $50 for the second-class cabin, and from $20 to $27 in the steerage. The winter rates are somewhat less, say, from $75 to $150 in the steamers *Lucania* and *Campania*, and from $60 to $150 in other fast boats. When the rush of travel is in full swing, say, from May to October, rooms must be secured months in advance. Tickets may then be held at a fictitious value, and those who will have *special* accommodation (suites of rooms, etc., etc.) must pay for it. A fellow-passenger with me, in one of the New York liners, not long since paid—so, at least, I was credibly informed—$3,000 for the single voyage for himself, his wife, two daughters, and two servants. The difference between an outside and an inside stateroom, in the busy season, may be $135 and upwards. At such times a room to yourself is a luxury that means money.

What about ocean steamers racing? The question was raised in the British House of Commons a few years ago, and elicited the answer that there is no law in the statute book forbidding it. Are not these ocean greyhounds built and subsidized with a special view to speed? Other things being equal, the fastest boat draws most passengers. A competing ship may be in sight or out of sight; it makes little difference. There is a race going on all the same, and the palm is awarded to the one that lands the mails in London or New York, as the case may be, in the fewest number of hours and minutes. Probably ninety-nine out of every hundred passengers on board the *Majestic* on a certain day in May, 1894, if placed in the witness-box, would swear that on that day an exciting race took place on the high seas, which ended in the SS. *Paris* outrunning the *Majestic*, and dashing across her bows in dangerous proximity! It was an optical delusion. Both ships, no doubt, were doing their level best, and had they continued their respective courses much longer, there is no saying what might have happened, but, at the proper time, Captain Parsell blew off steam, slowed his ship, put his helm down, and crossed the stern of the *Paris*. It was beautifully done.

And how about these so-called life-boats, hanging in the davits, so prettily painted, so neatly encased in canvas, and so firmly secured in their places? That they are useful sometimes, the writer knows from personal observation. On a recent voyage from Liverpool to New York we ran into a dense fog off the Banks of Newfoundland. The steam whistle gave forth its dolorous sounds all hours of the night, but the ship rushed on at her accustomed pace. At 4.20 a.m. most of us were awakened out of our slumbers by a violent shaking of the vessel. Had we been near land we might have fancied that the ship was grating along a pebbly bottom, but that could not be. Presently the engine stopped, and a loud roar of steam from the funnels brought most of the passengers on deck. It was a raw, damp morning, about daybreak, with fog as thick as burgoo all around. You couldn't see half the length of the ship. Everything on deck appeared to be at sixes and sevens. Where the after-boats had been ropes and tackles were swinging to the roll of the ship; orders were being given from the bridge in peremptory tones, a few sailors were hurrying here and there, yelling out their ready "Aye, aye, sir!" Down goes another boat. Three or four had already left the ship and disappeared in the mist. What is it all about? "Oh! we have run down a fishing schooner and smashed it to smithereens." Listen! voices of men in distress are heard; they shout louder and louder, and are answered, call for call, by the steam whistle. The ship had overshot the scene of the disaster, but was brought back to the spot by the instant reversal of her twin-screws—it was that that shook the ship as if it would have shaken her to pieces. The boats came in sight one by one, each to be greeted with a hearty cheer. Seven of the eight fishermen have been rescued! One had left the spar to which he had been clinging, thinking to swim for the ship, but he quickly went under and was seen no more. The longboat came first with two of the survivors; the life-boat came last, strange to say, full of water. She had struck a piece of wreckage and stove in her bow, but the men sat up to their waists in water—every sea washing over them—and plied their oars as merrily as though nothing had happened. They brought two of the fishermen, one of whom was too weak to grasp the rope ladder hanging over the ship's side, and was hoisted up by a cord passed round his body, a pitiful object. Reaching the deck they took him up tenderly and carried him below—to die in a few minutes. The remaining six, some of them badly bruised, were well cared for. A subscription on their behalf, added to the proceeds of a concert in the second cabin, realized about £380 sterling, which would cover the loss of their vessel and its cargo. The whole time occupied in the rescue was one hour and three-quarters. It was cleverly done: and the ship sailed on.

A fine instance of coolness and sound judgment in a sudden emergency has been related of Captain E. R. McKinstry, Lieut. R.N.R., of the SS. *Germanic*, which collided with the steamer *Cambrae* entering the Mersey in a dense fog. The *Germanic* had cut deeply into the broadside of the other ship, and filled the opening she had made like a wedge. Had the order been given to reverse the engine the result would have been disastrous, for the damaged ship must have filled and sank immediately, but with rare presence of mind the engines of the *Germanic* were kept moving slowly ahead, effectually preventing the rush of water until every soul on board was rescued. Captain McKinstry is a young man to have reached the top of his profession, and has already given many proofs of his gallantry and pluck. On several occasions he has risked his life to save that of others, notably during the naval review at Spithead, in 1887, when he jumped from the deck of the *Teutonic* to rescue a drowning sailor. Another instance of fine seamanship occurred recently on board the *City of Rome*, Atlantic liner, which had a narrow escape from destruction by fire on her voyage to New York with a large number of passengers on board. The coolness and skill of Captain Young on that occasion merited the highest praise. Mr. Wonham, of Montreal, one of the passengers, after describing the steps taken to subdue the flames, and to provide for the safety of the passengers and crew, concluded his narrative by saying, " I'm like the American who came to Montreal to enjoy a toboggan slide. He would not have missed the experience for a thousand dollars, but he wouldn't go through it again for ten thousand."

Leaving out of the count innumerable "tramps," there are many lines of steamships besides those already mentioned, keeping up regular sailings between Britain and United States ports. The Wilson Line, of Hull, has a fleet of about eighty steamers trading to all parts of the world, with weekly services from Hull and London to New York, and fortnightly from Newcastle and Antwerp. They also have a fortnightly service from Hull to Boston. The State Line, now incorporated with the Allan Line, has a weekly service from Glasgow to New York. The *State of Nebraska* and *State of California* are large and fine ships with excellent accommodation for passengers at low rates. The Atlantic Transport Line, with its fine fleet of twin-screw steamers, connects New York, Philadelphia and Baltimore with London every week. The North American Transport Company has also a numerous fleet plying between Norfolk, Va., and New York to Liverpool, Glasgow, Leith, Rotterdam and Hamburg. The Arrow Line runs from New York to Leith; the Manhanset Line, to Bristol and Swansea from New York. The Hill Line plies between London and New York, and the Lord Line between Baltimore and Belfast. The Chesapeake and Ohio Steamship Company sail their ships from Newport News and New York to London and Liverpool. The Blue Flag Line has regular communication with Baltimore and Glasgow, Liverpool, Dublin, Belfast and Rotterdam. The Lamport and Holt Line plies between New York, Liverpool and Manchester; the Bristol City Line weekly between New York and Bristol, while another line makes its terminus at Avonmouth. Barber & Co.'s steamers run regularly from New York to Leith, and from Norfolk, Va., and Newport News to Liverpool and Antwerp. The United States Shipping Company send their ships from Norfolk to Glasgow, Liverpool, Manchester, Leith and Hamburg.

Besides these there are many lines of steamships leaving New York at regular intervals for Bermuda, West Indies, Trinidad, New Orleans, South American ports, Mexico, Central America and San Francisco, *via* the Isthmus of Panama.

Continental Lines.

The great volume of emigration from the continent of Europe, and especially from Germany, has developed a correspondingly large steamship passenger traffic. France and Germany have, for many years, vied with each other as well as with the British shipping companies, in providing accommodation suitable to the demand. The result is several fleets of magnificent steamships little inferior in speed and luxurious appointments to the British and American lines.

The Hamburg–American Packet Company,

established in 1847, is the oldest of the German lines, and has now attained large dimensions. It began with a small capital and a fleet of three sailing ships. The average of their westward voyages from Hamburg to New York was about forty days, and eastward about thirty days; and they were accounted among the fast ships of their day. In 1867 the company owned a fleet of ten large transatlantic steamers, several smaller craft, a considerable amount of real estate and a commodious dry-dock. In 1872 the fleet had increased to twenty-five steamers, and a regular weekly service was maintained between Hamburg and

New York. The operations of the company at this time also extended to the West Indies, South America and Mexico: but 1888 was the *annus mirabilis* in the company's history, for it was then that a new departure was made, by the construction of twin-screw steamers destined to rival in speed and elegance the finest steamships afloat. In 1895 the company

THE "NORMANNIA," 1890.

owned a fleet of seventy ocean steamers and fifty-one river steamers, having a combined tonnage of 339,161 tons. Among its steamers there are no less than eighteen twin-screw passenger ships, all employed in the New York service. The four express boats of the line at present are the *Fürst Bismarck, Normannia, Augusta Victoria* and *Columbia*, all twin-screw ships of from 7,578 tons and 13,000 horse-power, to 8,874 tons and 16,000 horse-power.* Two of these were built at Stettin, Prussia, one at Birkenhead, and one, the *Normannia*, by John Elder & Co., on the Clyde. They have also a fleet of five large twin-screw steamers, especially adapted for live stock and fresh meat. In ten years, from 1881 to 1891, the Hamburg-American Line conveyed 525,900 passengers to New York, which was 50 per cent. more than either the Cunard or White Star Lines during the same period. The capital of the company is about $7,000,000, and its affairs are said to be exceedingly well managed. It has paddled its own canoe without State aid from the commencement, the only addition to its freight and passenger revenues being a moderate compensation from the American Government for carrying the mails from New York to Hamburg. The amount received for that service in 1896 was $30,030.75, being at the rate of about 44 cents per pound for letters and post cards, and $4\frac{1}{2}$ cents per pound for other postal matter.† The company is said to have in its employment a permanent staff of six thousand employees.

The *Augusta Victoria*, on her first voyage, made the fastest maiden trip then on record between Southampton and New York—7 days, 2 hours, 30 minutes. She has since made the run in 6 days, 19 hours, 19 minutes. The *Normannia* has done it in

---

* This was written before the Hispano-American war began; since then several of these vessels have been employed by the United States Government with a change of nomenclature.

† "U. S. A. Report on Navigation for 1896," p. 104.

6 days, 10 hours, 45 minutes, and the *Fürst Bismarck* in a few minutes' less time. The *Normannia*, built in 1890, was at that time claimed to be one of the finest steamships afloat. She is 520 feet long and 59 feet wide. On her trial trip she showed a speed of twenty-one knots. In addition to her main triple expansion engines, she makes use of fifty-six auxiliary ones, and is provided with a deck boiler, by

"AUGUSTA VICTORIA."

which steam is secured for her pumps in case of the main boilers being rendered useless by such an accident as befell the *Paris* a few years ago. Her passenger accommodation is unsurpassed. The music room is described as a "marvel of elegance." The decorations throughout are by the best European artists.

The line has not been exempt from marine disasters and loss of lives. The *Austria* was burned in 1858, when only sixty-seven were saved of the whole ship's company of 538. By the wreck of the *Schiller* on the Scilly Islands, in 1875, 331 persons perished. In 1883 the *Cimbria* was sunk off the coast of Holland, with the loss of 389 persons. The *Normannia*, on a recent trip, narrowly escaped collision with a huge iceberg, but thanks to her good "look-out" and her twin screws, she sheered off from the towering monarch just in time.

This company has recently added to its fleet one of the largest freight-carrying steamers afloat. The *Pennsylvania*, built and engined by Harland & Wolff, Belfast, has a carrying capacity of 21,762 tons, with accommodation for 200 first-class and 1,500 steerage passengers. Her length is 585 feet; breadth, 62 feet; draught of water when fully loaded, 30 feet. She has two balanced quadruple expansion engines, with five boilers, and carries a working pressure of 210 pounds of steam. Her three-bladed twin screws, each weighing $9\frac{1}{2}$ tons, make 76 revolutions per minute, developing a speed of fifteen knots an hour. The *Pennsylvania* left New York on her first voyage with a cargo of 18,500 tons measurement, said to be the largest cargo ever taken out of New York in one ship, if not the greatest that any ship in any part of the world has ever carried.

THE NORTH GERMAN LLOYD COMPANY.

This company, founded in 1857, has its headquarters at Bremen, and is also a very large concern, owning a fleet of eighty steamships, with a total tonnage of over

"PENNSYLVANIA," HAMBURG-AMERICAN LINE.
The largest cargo steamer afloat.

225,000 and 200,000 indicated horse-power. Among these are a number of very fine express steamers, mostly Clyde-built and fitted up with all the latest improvements in machinery and decoration. The *Kaiser Wilhelm II.*, the *Havel, Spree, Lahn, Trave* and *Fulda* are all well-known and favourite ships on the Atlantic route. Besides maintaining a weekly service between Southampton and New York, this company has a regular line running direct from New York to Genoa, Naples, Alexandria and other Mediterranean ports, and also lines running to India, China, Japan and Australia. A sad disaster was that which overtook the *Elbe* of this line in January, 1895, when she was struck amidships by a trading steamer, the *Crathie*, and sank in a few minutes, with the loss of 332 lives, only twenty-seven of the whole ship's company being saved. In December, 1896, the *Salier*, of this line, while on her voyage from Bremen to Buenos Ayres, foundered off the coast of Spain, when every soul on board perished, numbering about three hundred persons.

Eight gigantic steamships are being added to the already numerous fleet. Some of these have already been launched at Stettin, Germany. The largest of these leviathans is the *Kaiser Wilhelm der Grosse*, which arrived in New York on September 26th, 1897, having made her maiden voyage from Southampton in 5 days, 22 hours, 45 minutes, the fastest on record. Her average speed was over twenty-one knots an hour, and her daily runs as follows: 208, 531, 495, 512, 554, 564, 186; the total distance run was 3,050

"KAISER WILHELM DER GROSSE," NORTH GERMAN LLOYD LINE.

The largest passenger steamer afloat; holds the Blue Ribbon for the fastest voyage from Southampton to New York, the highest average speed, and the greatest day's run.

knots. Not only has the biggest ship beaten the Southampton record, but on her maiden trip she has made the fastest single day's run. This she did on the nautical day ending at noon on the 26th, when she reeled off 564 knots. At times she developed twenty-two knots. Her coal consumption, however, was heavy, being nearly five hundred tons a day. She was commanded by Captain H. Englebart. Her return voyage to Plymouth was made in 5 days, 15 hours, 10 minutes; her average speed was about 21.40 knots, and her daily runs were 367, 504, 500, 507, 510, 519, 55; total, 2,962 knots.*

The *Kaiser der Grosse* is 649 feet in length, 66 feet in width, and 43 feet in depth. She is rated at 14,000 tons burthen and 30,000 horse-power. She has quadruple expansion engines, working at a steam pressure of 213 lbs., and turning her mammoth twin screws at the rate of seventy-seven revolutions per minute, and is otherwise conspicuous by her four funnels. Even the *Pennsylvania* is thrown into the shade by this new-comer. She is designed to carry 20,000 tons of cargo, and from 1,500 to 2,300 passengers. She is the largest steamship afloat at the present time, having larger carrying capacity than the famous *Great Eastern;* but her supremacy will be short-lived, for the new *Oceanic,* of the White Star Line, is still larger, and may prove faster. To

---

* Last April the great *Kaiser* surpassed her previous record, making the voyage from New York to Southampton (3,065 knots) in 5 days, 17 hours, 8 minutes, showing an average speed of 22.35 knots per hour.

load this great ship entirely with wheat would require the produce of a field of 40,000 acres, at sixteen bushels to the acre; and to supply her full complement of passengers would depopulate a good-sized town. The *Kaiser* is essentially a new type of ocean steamship—a magnificent experiment, which will be watched with great interest in shipping circles everywhere, and one that is not unlikely to set the fashion for ships of the next decade.

The Compagnie Générale Transatlantique,

commonly known as the French Line, entered the lists of competition in 1862, and has developed into a first-class marine service. The early ships of this company were iron paddle-wheel steamers, which were built by Scott & Company, of Greenock, but, owing to the prohibitory duty imposed on foreign-built vessels, it was found to be more advantageous to have them built in France, the more so as the Government had introduced the system of giving large "construction bounties." This French company has now a magnificent fleet, comprising upwards of sixty steamships. The Atlantic service employs six very fine express steamers, *La Touraine*, *La Bourgogne*,* *La Bretagne*, *La Champagne*, *La*

---

* The "Bourgogne" Disaster.—Since the sinking of the *Eutopia* in Gibraltar Bay in 1891, no such marine disaster has occurred as that which recently befell the SS. *Bourgogne*—a tragedy in some respects the most appalling that has ever been recorded. This vessel of 7,795 tons—one of the finest of the French line of steamers—sailed from New York for Havre on the

Gascogne, *La Normandie*, all of them built in France except the last named, which was built at Barrow-in-Furness, in 1882. The *Touraine* was built at the company's building yard, at St. Nazaire, in 1890. She is a steel twin-screw ship of 10,000 tons net, and 14,000 horse-power. Her length is 520 feet, breadth 56 feet, and depth 34½ feet. She has triple expansion engines, and is classed as a nineteen-knot boat. She has made the voyage from Havre to Sandy Hook (in July, 1892) in 6 days, 17 hours, 30 minutes, the fastest on record between these ports, the average speed being 19.63 knots,

2nd of July, 1898, with a ship's company, including passengers and crew, of 726 souls. Early on the morning of the 4th, when about sixty miles south of Sable Island, during a dense fog, and while running at the rate of some eighteen knots an hour, she came into collision with the British sailing ship *Cromartyshire*, of 1,554 tons, and in a very short time foundered, carrying down with her about 520 persons. Had it not been for her collision bulkhead the *Cromartyshire* must have sunk, too. As it was, she was badly damaged, but hove to all day in the hope of picking up survivors. In the meantime the Allan SS. *Grecian* came up to the scene of the disaster, the rescued passengers were taken on board, and the disabled ship was towed into Halifax harbour. The survivors were the purser of the steamship, three engineers, thirty of the crew, and 170 passengers—204 in all. Of the seventy-two ladies in the first cabin only one was saved. Captain Deloncle, commander of the *Bourgogne*, was a lieutenant in the navy, and a knight of the Legion of Honour, having under him a competent staff of officers who appear to have done what they could to save the lives of others. All of them went down with their ship into the sailor's grave. The loss of life was appalling, but even more heartrending were the accounts given of the barbarous conduct of some of the steerage passengers and sailors in the terrible struggle for self-preservation.

and the best day's run, 501 knots. The company's capital is said to be $8,000,000, and its credit is good. The line is largely subsidized by the French Government, and receives compensation from the United States for carrying the mails from New York to Havre, the amount thus received in 1896 being $32.806.86. Until the loss of the *Bourgogne*, the most serious disaster that had overtaken the line was the wreck of the *Ville de Havre*, in November, 1873, from collision with an iron sailing ship, the *Lochearn*, which involved the loss of 226 lives, eighty-seven being rescued. Besides the American business, which is very large, the company has extensive trade connections with the Mediterranean and the West Indies.

THE NETHERLANDS LINE,

officially styled the "Nederlandsch-Amerikaansche Stoomvart Maatschappig," of Rotterdam, has a fleet of thirteen steamers, most of them from the shipyard of Harland & Wolff, Belfast, and ranging from 3,000 to 4,000 tons each. They are very fine boats of their class, and have attracted a fair share of the passenger traffic between New York and Amsterdam and Rotterdam, sailing alternately for these ports every week, calling at Boulogne-sur-mere. They carry the United States mails, which do not seem to be very weighty, as the *pay* only amounted to $165.03 in 1896. The latest addition to the fleet is the *Spaarndam*, formerly of the White Star Line (the *Arabic*), a fifteen-knot ship, of 4,368 tons and

3,000 horse-power. The company, which commenced this business in 1872, has a capital of $1,680,000.

THE THINGVALLA LINE,

dating from 1879, is a Danish enterprise, with a regular service between Copenhagen and New York, consisting of five ships, the largest of which is the *Amerika*, of 3,867 tons, formerly the *Celtic*, purchased from the White Star Line in 1893. This line came into notoriety in 1889 through the foundering of one of their vessels, the *Danmark*, in mid-ocean. She had on board 735 souls. On April 5th she was sighted by the British steamship *Missouri*, Captain Hamilton Murrell. On April 6th, though a heavy sea was running, by an act of heroism almost unparalleled, Captain Murrell threw some of his cargo overboard, and in four and a half hours saved every soul by means of boats and lines, landing some at St. Michael's, Azores, and the rest at Philadelphia. The gallant rescue was suitably acknowledged by public testimonials from Britain and America to the captain, his officers and crew.*

* Fry's "History," p. 309.

CHAPTER V.

STEAM TO INDIA AND THE EAST.

DURING the earlier years of commerce with India, the route from Britain was by the Mediterranean, the Black Sea, and the Caspian, through Persia, reaching India at its northern extremity. The sea route, *via* the Cape of Good Hope, was discovered by the Portuguese in 1497, and continued to be the great highway of commerce to the East until our own times. Although circuitous, the Cape route was infinitely preferable to that of inland seas and deserts infested by hostile tribes, to say nothing of the advantage of reaching destinations without transhipment.

The importance of India as a field of British enterprise began with the incorporation of the East India Company in the year 1600. From a small trading company it gradually became a vast aggressive monopoly, with a large standing army at its back, and a numerous fleet of ships that served the double purpose of carrying merchandise and fighting the French, or any other covetous enemy. In 1811, when the company had reached the zenith of its power, it owned sixty-seven ships, each armed with from 30 to 38 guns; thirty-one ships of from 20 to 28 guns, and fifty-two ships of from 10 to 19 guns. The sea route to Calcutta was over 13,000 miles, and not unfrequently a whole year was occupied in making the round trip. In the days of clipper ships, however, the single voyage was sometimes accomplished inside of one hundred days.

Lieutenant Thomas Waghorn, R.N., an English

THE CAMEL-POST—"SHIP OF THE DESERT."

naval officer, applied to the British Government for assistance in carrying out a project he had conceived of opening communication by steam between Britain and her great East Indian Empire. The result of his labours was the opening up of the overland mail route, as it was called, consisting at first of a steam service from Marseilles to Alexandria, thence by camel and Nile steamer to Cairo, a caravan across the desert to Suez, and steamers *via* the Red Sea to Bombay and Calcutta. The next improvement was the substitution of a railway for "The Ship of the Desert," in 1858, and the transmission of the English mails to Brindisi instead of Marseilles, and finally, the construction of the Suez Canal by Ferdinand Lesseps, the French engineer, at a cost of sixty million dollars. The canal is ninety-nine miles long with a width of 327 feet for 77 miles and 196 feet for the remaining 22 miles; the depth was originally twenty-six feet throughout, but the canal is undergoing progressive enlargement and deepening. The British Government in 1875 acquired by purchase shares in the enterprise to the value of £4,000,000 sterling. By a convention signed in 1888, the canal was exempted from blockade, and vessels of all nations, whether armed or not, may pass through it in peace or in war.* The North German Lloyd SS. *Frederick the Great*, of 10,500 tons register, which passed through the canal a few months ago *en route* for Australia, is the largest vessel that has passed through it. The canal was first opened for traffic in 1869.

By the overland route the distance from London to Bombay has been reduced to 5,221 miles, and to Calcutta, 6,471 miles. The contract time for the transmission of mails is 16½ and 18½ days respectively. Sir Douglas Fox, engineer of the railway from Acre to Damascus, speaking of the proposal to extend that road to the mouth of the Persian Gulf, prophesied that in a few years the journey from Charing Cross to India will be covered in eight days! It will be

* "Whitaker's Almanack," 1897, p. 543.

accomplished in about the same length of time, *via* Russia, when the great trans-Siberian railway is completed. When that is accomplished, the actual running time of an excursion around the world may possibly be reduced to thirty days or even less.

In preceding pages reference has been almost exclusively made to the development of steam navigation on the North Atlantic; a brief allusion must now be made to the effects produced on the commerce of other parts of the world by the introduction of steam power. The Atlantic steamers were probably the first to bridge the ocean; they are, perhaps, the most numerous to-day; certainly they include some of the largest and most magnificent specimens of marine architecture in existence, but they are only a wing of the world's fleet of steamships. There are other great lines of ocean steamers performing services of equal importance elsewhere, though with their history and their "records" we are less familiar. An excellent summary of the lines of communication with India, and the East generally, is given in "Whitaker's Almanacks" for 1896 and 1897, under the caption of "Our Ocean Mail." Mr. Macdonald, in "Our Ocean Railways," devotes a couple of chapters to an interesting survey of this branch of our subject.

THE PENINSULAR AND ORIENTAL COMPANY, commonly known as the "P. & O." Co., is the second oldest steam-packet company in existence. It had its origin in a small steamship undertaking, started in

1836 under the name of the "Peninsular Company," to trade between Falmouth and Lisbon. Their first vessel was the *William Fawcett*, a paddle-steamer of 206 tons, built in 1829. The first steamer despatched for India by this company was the *Hindostan* of 1,800 tons and 250 horse-power, about the year 1842. From that time until now the history of the company has been a continuous record of progress and pros-

P. & O. STEAMSHIP "CALEDONIA."

perity. They now carry the mails not only to India, but to China and Australia, having in their service a magnificent fleet of over sixty steamers, ranging from 2,500 tons to 7,560 tons, and aggregating some 220,000 tons. The SS. *Caledonia* is at present the largest and fastest vessel employed in the Indian trade, and has succeeded in landing her mails in Bombay within 12½ days of their despatch from London. Their contract time for the delivery of mails in Shanghai is

37½ days, and 35½ days to Melbourne, Australia. Over $35,000,000 have been expended on the fleet of the P. & O. Company in the last twenty years, and they are now building several steamers of 8,000 tons for the mail service. Among the larger boats of the fleet at present are the *Arcadia*, 6,670 tons; *Australia*, 6,901; *Himalaya*, 6,898; *Oceanea*, 6,670, and the *Victoria*, 6,527 tons. During the Crimean war, and at the time of the Indian mutiny, this company rendered important services to the Government in the rapid conveyance of troops and stores. The regularity with which the mail service has been conducted is remarkable when the length of the routes is considered. It is seldom that the mails are even an hour late in being delivered. The ships combine all the latest improvements in their construction, machinery and internal fittings.

The P. & O. steamers leave London every Saturday for India, and fortnightly for Australia and China. The first-class ordinary fare to Bombay, Madras, or Calcutta by this line is £55 sterling; second-class, from £35 to £37 10s. To Adelaide, Melbourne or Sydney, Australia, first-class, £60 to £70; second-class, £35 to £40. To China and Japan, first-class, £73 10s.; second-class, £42. The rates for special accommodation are, of course, considerably higher.

The Orient Steam Navigation Company was formed in 1877 by two well-known shipping firms—Anderson, Anderson & Co. and F. Green & Co. The first steamer to leave London under the flag of the Orient Line was the *Garonne,* acquired by purchase, and followed by the *Chimborazo, Lusitania* and *Cuzco.* Two of these are now used on exclusively pleasure cruising voyages in the Mediterranean and elsewhere, while a number of large and powerful ships have been built for the mail line. The *Orient,* built by Robert Napier & Sons, Glasgow, in 1879, was the largest steamer constructed on the Clyde up to that time. She was 400 feet long, 5,365 tons register, and with engines of 6,000 indicated horse-power. Her speed was seventeen knots on her trial trip. The latest additions to the fleet are the *Ophir,* 6,057 tons; *Orizaba,* 6,077; *Oroya,* 6,057, and the *Ormuz,* 6,031 tons. The *Ophir* is 482 feet long, 53 feet beam, and 37 feet moulded depth. She is fitted with triple expansion engines and twin screws, and all the other modern improvements which go to make up a "floating palace." The company receives a subsidy from the Imperial Government of £85,000 sterling per annum for carrying the mails, which are despatched fortnightly from London calling at Plymouth, Gibraltar, Naples, Port Said, Suez, Colombo, Albany, Adelaide, Melbourne, and Sydney, Australia.

The British India Steam Navigation Company dates from .1855, when the East India Company first took steps to establish a mail service between Calcutta and Burmah. In 1862 the name was changed from the Calcutta and Burmah Steam Navigation Company to that which it now bears. Since then the business has greatly increased, and it now boasts of having more steamers than any company trading to the East. Its fleet consists of 106 vessels with a total tonnage of about 270,000. They are nearly all called by Eastern names, such as the *Golconda*, 6,036 tons; *Matiana*, 5,000 tons; *Okhla*, 5,283 tons; *Onda*, 5,272 tons, and *Obra*, 5,456 tons. The distance annually travelled by ships of this line counts up to 5,000,000 miles. The sailings are about fortnightly from London to Colombo, Madras and Calcutta. The fares to Madras and Calcutta are from £47 10s. to £52 10s., according to accommodation. The first steamers of the line—the *Cape of Good Hope* and the *Baltic*—were despatched to India *via* the Cape. The *India* of this line is said to have been the first steamer to pass through the Suez Canal. In 1872 a contract was entered into with the East India Company for a monthly service from Aden to Zanzibar. Then a coast line was established from Bombay to Calcutta, calling at eighteen intermediate ports, with a branch line running up the Persian Gulf. In 1880 arrangements were made with the Government of Queensland for a mail service that soon developed into a large trade. At the breaking out of the mutiny in 1857, a detachment of the 35th Regiment was brought up from Ceylon to Calcutta by one of the ships of this line most opportunely. Again, in 1863, thirteen steamers of this fleet were taken up by the Government in connection with the Abyssinian expedition.

Some years ago the *Quetta*, of this line, on her voyage from Queensland, struck a rock in Torres

Straits and sank in a few minutes with the loss of 133 lives. Among the survivors was a plucky young lady, a Miss Lacy, who, after having spent twelve hours upon a raft, attempted to swim ashore, and kept afloat in the water for twenty-four hours without a life-belt or support of any kind, until she was picked up by a boat from a passing steamer.

The Clan Line, established in 1878, has a fleet of

THE "QUETTA" GOING UNDER, 1890.

some thirty-five ships, all rejoicing in the prefix of "Clan" to their names. They are comparatively small vessels, the largest of them being the *Clan Grant*, 3,545 tons; *Clan MacArthur*, 3,934; *Clan MacIntosh*, 3,985; *Clan MacPherson*, 3,921, and *Clan Matheson*, 3,917 tons. They run from Glasgow and Liverpool to Bombay; from the same ports to Colombo, Madras and Calcutta; also to Cape Colony and Natal, Delagoa Bay, Beira and Mauritius. The saloon fare by this line from Liverpool to Madras or Calcutta is £45; second class, £30.

The Bibby Line has long been famous on the Mediterranean. It is now the direct route to Burmah, and controls a large share of the trade with Ceylon and southern India. It employs five of Harland & Wolff's first-class steamships—the *Staffordshire*, *Shropshire* and *Cheshire*, twin-screw ships of 6,000 tons; and the *Lancashire* and *Yorkshire* of 4,260 tons each. This line is the recognized route for officers returning from India at the expiry of furlough. The sailings are from Liverpool to Egypt, Colombo, southern India and Rangoon. Only first-class passengers are carried. Fare to Rangoon, £50.

The Shaw, Savill & Albion Company, formed some thirteen years ago, has been very successful. It has five fast mail steamers—the *Arawa*, 5,026 tons; *Doric*, 4,786; *Ionic*, 4,753; *Tainui*, 5,031, and the *Gothic*, 7,730. Besides these they have a large number of cargo steamers and sailing ships. The *Gothic* is said to be the largest steamship employed in the Australian trade, and the *Arawa* the fastest, having made the run from Plymouth to New Zealand in 38 days, 30 minutes; and from New Zealand to Plymouth in 35 days, 3 hours, 40 minutes—the fastest on record.

The Union Steamship Company of New Zealand advertises to take passengers from Auckland to England, *via* San Francisco, in *thirty-one* days! Saloon fare, £66; steerage, £32 11s. 7d.

The Anchor Line has two services to India: (1) from Liverpool to Bombay and Kurrachee; (2) from Liverpool to Calcutta. The sailings in each case are about once a fortnight. Though chiefly adapted for freight, they carry a considerable number of passengers at low rates, say, to Bombay or Calcutta, first-class, £45, and second-class, £30. The City Line has also two distinct services, the same as the Anchor Line, to Bombay and Kurrachee and to Calcutta. The fares are the same. This line has a fleet of fourteen steamers, among the largest of which are the *City of Bombay*, 4,548 tons; *City of Vienna*, 4,672 tons; *City of Oxford*, 4,019 tons; *City of Calcutta*, 3,906 tons.

The Hall Line, from Liverpool to Kurrachee and to Bombay, calling at Marseilles, sails about once in three weeks The ships are all about four thousand tons. The fare from Liverpool to Bombay is, for first-class, £47 10s., and for second-class, £30. The Henderson Line has sailings from Liverpool to Rangoon every three weeks, with accommodation for second-class passengers. The New Zealand Shipping Company has a fine fleet of steamers, from four thousand to six thousand tons, sailing once in three weeks from London to New Zealand ports, Tasmania and Australia. Fare to Auckland, £68, and to Melbourne or Sydney, £72. The North German Lloyd Line has a monthly service from Southampton to China and Japan, and also to Australia. Holt's Line has sailings once a fortnight to China, Japan and Australia from Liverpool.

There are various other lines of steamers in the Eastern trade, but the above-named are the most important, unless we include the *Messageries Mari-*

time and the *Rubattino* Lines, both of which are formidable competitors for the freight and passenger traffic. The former is a French line, which has been in existence since 1852, and has attained a high rank. The fleet numbers about sixty vessels, many of them very large, handsomely fitted and fast. They are noted for their elaborate *cuisine*, which attracts a certain class of travellers, and though their rates are somewhat higher than the other first-class lines, they have long been very popular. The line to India has sailings from Marseilles and Trieste once a fortnight. The Messageries Company receives a very large subsidy from the French Government. The *Ville de la Ciotat*, built for the Australian trade, is a magnificent ship of 6,500 tons and 7,000 horse-power. The *Rubattino* is an Italian line, which has a numerous fleet of steamers, chiefly adapted for the Mediterranean trade; but they have also a number of large vessels sailing at regular intervals from Genoa and Naples to Bombay.

The Eastern trade is enormous. The total exports from and to India, Ceylon, the Straits, Labuan and Hong Kong amounted in 1889 to $1,031,000,000. The exports and imports to and from Australia amounted in the same year to nearly $526,000,000.* The net tonnage which passed through the Suez Canal in 1894 was 8,039,105 tons.

* "Our Ocean Railways," p. 119.

Steamship Lines to Africa.

The African Steamship Company is one of the oldest and largest shipping concerns in the African trade. It originated in 1832 as a private expedition by MacGregor Laird, of Liverpool, for the purpose of exploring the Niger River. In 1852 the company received its charter, and agreed to perform a monthly mail and passenger service to West Africa in consideration of an annual subsidy of £30,000. The pioneer ships were the *Forerunner, Faith, Hope* and *Charity*. Year after year numerous fine vessels were added to the fleet, among which are the *Leopoldville*, 3,500 tons; *Assaye*, 4,296 tons; *Mohawk*, 5,658 tons, and the *Mobile*, 5,780 tons. In 1891 this company amalgamated with the Elder, Dempster Company of Liverpool, and now have regular services from Liverpool to South-West Africa; from Hamburg and Rotterdam to West and South-West Africa; and from Antwerp to South-West Africa.

The Union Steamship Company was first formed in 1853, with a fleet of five small collier steamers. In 1857 a contract was obtained for a mail service to the Cape for five years at £30,000 a year. The service proved so satisfactory that the contract was renewed and extended. The Union Line now carries the English mail to the Cape and Natal, and also from Hamburg, Rotterdam, Antwerp and Southampton to Cape Town, Port Elizabeth, East London and Natal, making calls at Madeira and Teneriffe. The *Scot*, built for this company by the Dennys of Dumbarton, is a fine ship of 6,850 tons, and has made the shortest voyage on record from Southampton to Cape Town, viz., 14 days, 11 hours. The *Norman*, of 7,537 tons, one of Harland & Wolff's steel twin-screw ships, is the largest vessel employed in the South African trade. The *Guelph*, *Greek*, *Gaul* and *Goth* are also twin-screw ships, close upon 5,000 tons each.

The Castle Line, founded by Sir Donald Currie in 1872, has attained a front rank in shipping circles. Since 1876 this line has carried the Royal mails between England and South Africa. The fleet numbers some fourteen or fifteen powerful steamers, of from 3,600 to 5,636 tons, such as the *Tantallon Castle*, *Dunottar Castle*, *Roslin Castle*, *Doune Castle*, etc. The voyage to the Cape of Good Hope, which used to occupy from thirty to thirty-four days, is now accomplished by the Castle Line in half that time. Until recently this company enjoyed an enviable immunity from marine disasters, not having lost a single life through mishap of any kind; but one dark and hazy night in June, 1896, one of the best-known ships of the line—the *Drummond Castle*—while attempting to sail through the perilous channel between the Island of Ushant and the mainland, struck a sunken rock, and almost immediately went to pieces, only three persons out of a ship's company of 250 having survived to tell the tale.

The British and African Steam Navigation Company, established in 1868, conveys passengers and mails from Liverpool to the west coast of Africa. It has a fleet of twenty-four steamers, and maintains seven distinct services. It is under the management of Elder, Dempster & Co. The ships are from 2,000 to 3,000 tons register, and derive their names from the rivers and ports which they frequent, *e.g.*, the *Bakana, Batanga, Loanda, Boma, Calabar*, etc.

The Natal Line, from London to Natal, Delagoa Bay, and other East African ports, was founded by Messrs. Ballard, King & Co. in 1879. They employ a fleet of ten steamers, ranging from 1,600 to 2,750 tons—larger vessels being unable to cross the bar at Natal. They have also a colonial service under contract with the Government of Natal, from Cape Colony and Natal to Madras and Calcutta. There is also the Aberdeen Line from London to Natal direct; the British and Colonial Steam Navigation Company from London to South and East Africa; the East African mail service of the British India Line, and the German East African Line. The fares from London to Delagoa Bay vary according to the class of ships, from 35 guineas by the Natal Line, to £67 10s. by the British India Line.

West Indies and Pacific Lines.

The Royal Mail Steam Packet Company, the ships of which line sail from Southampton to the West Indies, Central America, North and South Pacific, Brazil and River Plate, was founded in 1839, and has a large fleet of powerful steamers.

The *Danube, Nile, Clyde, Thames, Magdalena* and *Atrato* are all over 5,000 tons, with engines of from 6,773 to 7,500 indicated horse-power. Among the smaller ships is the *Trent*, a namesake of the historic vessel which was boarded by the United States cruiser, *San Jacinto*, in 1861, when the seizure of Slidell and Mason nearly provoked a war with Great Britain. The West India and Pacific Steamship Company, with a fleet of seventeen steamers, keeps up a good line of communication between Liverpool, the West Indies, the Gulf of Mexico, and the Caribbean seas. The *American* and *European* are each 7,730 tons; the *Barbadian, Cuban, Jamaican, Mexican* and *Tampican* are from 4,020 to 4,500 tons.

The Pacific Steam Navigation Company, incorporated in 1840, conducts a line of mail steamers from Liverpool to Brazil and River Plate, continuing the voyage to the west coast of America *via* the Straits of Magellan. This company are the pioneers of steam navigation along the southern shores of the Pacific, and between Europe and the West Coast. They have also running in the Orient Line, from London to Australia, four of their largest steamers, viz., *Orizaba, Oroya, Oruba* and *Orotava*, all over 6,000 tons. They have a large fleet of other ships, such as the *Orissa, Orcana, Potosi, Liguria, Iberia*, ranging from 4,000 to 5,000 tons each, and they are building others of large dimensions.

Messrs. Lamport and Holt have a fine fleet, consisting of over sixty steamers, running from Glasgow,

Liverpool, Manchester, London, Antwerp and New York, to Brazil, River Plate, and the west coast of South America. A large percentage of their steamers are capable of carrying between 5,000 and 6,000 tons of cargo, and have a speed of from $10\frac{1}{2}$ to 12 knots at sea. They also carry a limited number of passengers. The largest of their steamers are the *Canova*, 5,000 tons; *Cavour*, 5,500 tons; *Cervantes*, 5,000 tons, and the *Horace*, 4,000 tons. The Wilson Line—Thomas Wilson, Sons & Co. (Limited), Hull—in addition to their North American lines of steamers, have a fortnightly service to Bombay and Kurrachee, a monthly service to Australia, and a line of steamers running to River Plate ports, to suit the trade.

The fare from Southampton to the West India Islands runs from £25 to £35; from New York, by the Atlas Line, $50; and to Bermuda, by the Quebec Steamship Company, sailing from New York every Thursday, $25.

The Canadian Trans-Pacific Steamships.

The idea of connecting the Atlantic with the Pacific Ocean by a railway through British territory had long been a cherished vision of British and Canadian statesmen, railway engineers, and travellers in the far West; but owing to the vastness of such an enterprise for a people of four millions, a "baseless vision" it continued to be until after the confederation of the provinces in 1867. Twenty years before that time, Major Carmichael Smyth, writing to "Sam Slick," advocated the construction, by convict labour, of a trans-continental railway through British territory, and prepared a map on which the possible route of such a railway was marked—almost identical with that of the Canadian Pacific Railway.* Hon. Joseph Howe, in course of a speech made at Halifax in 1851, said he believed that many of his auditors would live to hear the whistle of the steam-engine in the passes of the Rockies, and to make the journey from Halifax to the Pacific in five or six days. Hon. Alexander Morris, in his lecture, "Nova Britannia," delivered in 1855, predicted the accomplishment of such an enterprise in the near future. Judge Haliburton, Sir Edward Bulwer, Sir George Simpson and other *savans* had all prophesied after the same manner. Sure enough, it was one of the earliest measures that came to be discussed in the first Parliament of the new Dominion. Preliminary surveys were commenced in 1871 by Sandford Fleming, chief engineer, and the work of construction by the Government followed soon after. But it early became apparent that Government machinery was ill adapted for successfully dealing with a work of such magnitude, and one unavoidably leading to political complications. It was therefore resolved to have the road built by contract. Finally, in 1881, the Canadian Pacific Railway Company was organized, the prime movers of the enterprise being Messrs. George Stephen and Donald A. Smith, of Montreal. At this time the

---

* "Statistical Year-Book, 1896," under Railways, p. 20.

Government had under construction 425 miles between Lake Superior and Winnipeg, and 213 miles in British Columbia. This company undertook to complete the railway from Quebec to Vancouver, a distance of 3,078 miles, within ten years, for which they were to receive $25,000,000 in money, and twenty-five million acres of land, together with the sections of railway already under construction by the Government, the entire railway when completed to remain the property of the company. Such was the energy of the contractors and the skill of their engineers, the railway was completed in one-half of the time stipulated; for on the 7th of November, 1885, the last rail was laid on the main line, and by next midsummer the whole of the vast system was fully equipped and in running order. The opening of the Canadian Pacific Railway was followed by an immense development of traffic.

The natural outcome of this was the inauguration of a line of steamships from the western terminus of the road to Japan and China, as well as to Australia. Sooner than might have been expected, three very fine twin-screw steel ships were built at Barrow-on-Furness for the Canadian Pacific Railway Company, under contract with the Imperial and Dominion Governments for carrying the mails to Japan and China. The ships are named the *Empress of India*, *Empress of China* and *Empress of Japan*.

The inauguration of the "Empress Line" was of the nature of a magnificent ovation. The maiden trips of the three sisters were largely advertised in connection with an all-the-way-around-the-world trip, *via* Gibraltar, Suez, Colombo, Hong Kong, Yokohama and Vancouver, and thence by the Canadian Pacific Railway across the continent and home again by any of the Atlantic liners, all for the modest sum of $600. The proposal took readily, with the result that the three ships had a full complement of cabin passengers, all of whom expressed themselves as delighted with the arrangements which had been made for their comfort. The first steamer, the *Empress of India*, with 141 saloon passengers, reached Hong Kong on the 23rd of March, 1891, under easy steam, in forty-three days from Liverpool; leaving Hong Kong on April 7th, she reached Yokohama on the 16th. She left on the 17th, and, although encountering a very heavy gale, reached Victoria, B.C., in 10 days, 14 hours, 34 minutes, an average speed of 406 miles a day, or just 17 knots an hour. The regular monthly service from Vancouver to Japan and China commenced in the autumn of the same year. For this service the company receives an annual subsidy of $300,000, and an additional subvention of about $35,585 to secure their services to the British Government whenever the vessels may be required as transports or cruisers. The three ships are all just alike. They are painted white and are beautiful models, with raking masts and funnels, and graceful overhanging bows. They are each 485 feet in length, 51 feet moulded breadth, and 36 feet in depth; gross tonnage about 6,000 tons each. They have triple expansion engines of 10,000 indicated horse-power, which with 89 revolutions per minute, and a con-

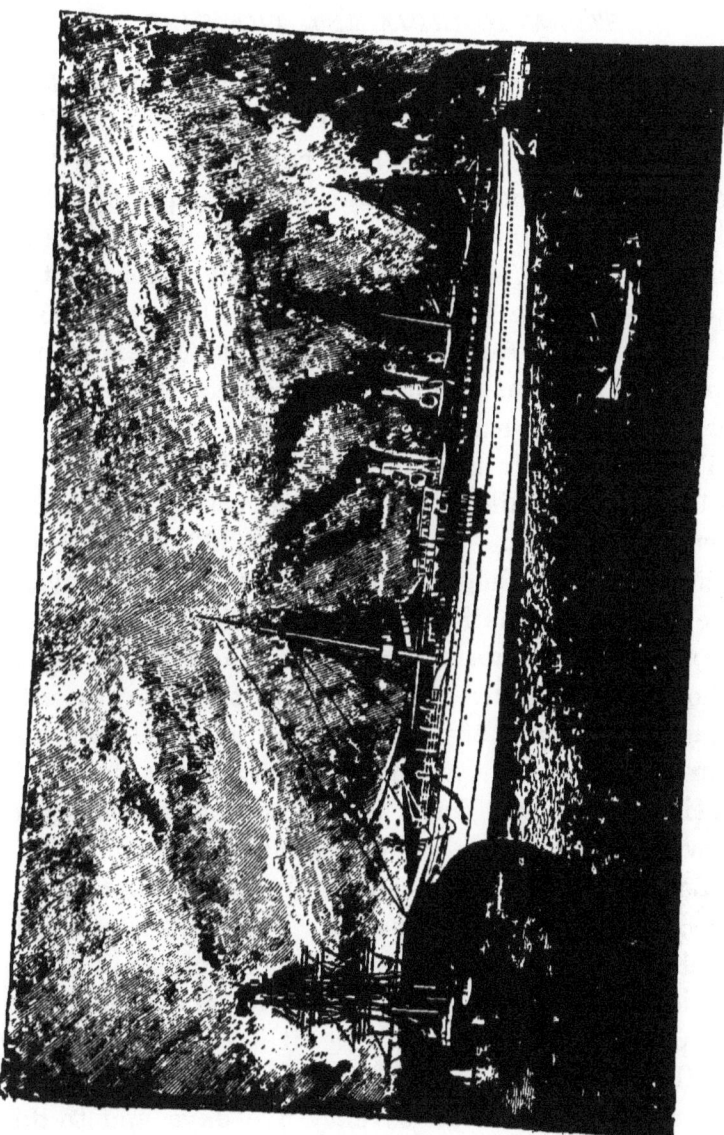

C. P. R. STEAMSHIP "EMPRESS OF JAPAN."

sumption of only 170 tons of coal a day, drive the ships at an average speed of 17 knots an hour. The arrangements and fittings for passengers are of the most complete and even luxurious description. The saloons and staterooms are tastefully decorated, handsomely furnished, and brilliantly lighted by electricity. They have ample accommodation for 180 first-class, 32 second-class, and 600 steerage passengers, with capacity for about 4,000 tons of cargo. They cost about $1,000,000 each.

The distance from Vancouver to Hong Kong is 6,140 nautical miles; the average passage is about twenty-two days. Yokohama is 4,300 knots from Vancouver, and the average passage is from eleven to eleven and a half days; but in August, 1891, the *Empress of Japan* made the voyage in 9 days, 9 hours, 39 minutes, the shortest time on record, being at the rate of eighteen and a half knots an hour. After a fairly quick railway run across the continent to New York, and close connection with a swift Atlantic greyhound, her mails were delivered in London in the unprecedentedly short time of 20 days, 9 hours from Yokohama. This feat astonished London, and gave rise to speculations of rapid communication with the East hitherto undreamed of. Even with existing facilities, it is now not only possible, but it is easy to go round the world by this route in less than seventy-five days, and to do it in palatial style for less than $1,000!

In connection with the Canadian Pacific Railway a line of steamers commenced a monthly service in

1893 between Vancouver and Australia, calling at Shanghai, Sandwich Islands, Brisbane, Queensland and Sydney, N. S. W. The pioneer ships are the *Warrimoo* and *Miowera*, of about 5,000 tons each, which have so far given a very satisfactory service. They receive a small subsidy from the Canadian and Australian Governments as a means of developing trade and commerce between the two countries, and as forging another link in the chain that binds the colonies to the Mother Country. A third steamer, the *Aorangi*, has recently been added to this line. The distance from Vancouver to Sydney, direct, is 6,832 knots, and the voyage has been made by the *Miowera* in $19\frac{1}{2}$ days, showing that with a fast Atlantic service and close connections the quickest route from England to Australia will be *via* Canada.

Still more recently, the unprecedented rush of adventurous gold-seekers to the Klondike has induced the Canadian Pacific Railway Company to inaugurate another line of steamships to ply between Victoria and Vancouver and ports on the northern Pacific coast. Two very fine Clyde-built steamers have been placed on this route, the *Tartar* and the *Athenian*, of 4,425 and 3,882 tons, respectively. These vessels are fitted up in first-class style, with excellent accommodation for large numbers of passengers. With the exception of the Empress Line of steamships to Japan and China, they are said to be much the finest steamers on the North Pacific coast.

George Stephen, now Lord MountStephen, was born at Dufftown, Banffshire, Scotland, June 5th, 1829 : came to this country in 1850, when he entered into business in Montreal, and was the pioneer of the woollen manufacturers in Canada. He became President of the Bank of Montreal and also of the Canadian Pacific Railway, which was completed mainly through his Lordship's energy. Sir George Stephen, Baronet —so created in January, 1886—was elevated to the British peerage in May, 1891.

Donald A. Smith, now Lord Strathcona and Mount Royal, who was associated with Lord MountStephen in the construction of the Canadian Pacific Railway, was born at Archieston, Morayshire, August 6th, 1820. He came to Canada in 1839 on the Hudson's Bay Company's staff, and eventually became Governor of that corporation. He has represented the city of Montreal in the Dominion Parliament, is President of the Bank of Montreal, and Chancellor of McGill University. He succeeded Sir Charles Tupper as High Commissioner for Canada in London in August, 1896. He received the honour of knighthood from Her Majesty the Queen in May, 1886, and was raised to the peerage on the occasion of Her Majesty's Diamond Jubilee in 1897. The gifts of both these gentlemen for educational and philanthropic purposes have been upon a princely scale, running up into millions of dollars.

CHAPTER VI.

STEAM IN THE BRITISH NAVY.

The British Navy—Marine Distances—Sunday at Sea—Icebergs and Tidal Waves.

GREAT as have been the changes brought about by steam navigation applied to commercial uses, the transformations of the navies of the world have been even more remarkable. It seems almost incredible that at the commencement of Her Majesty's reign there were less than twenty steamships in the British navy, and none of them over 1,000 tons burthen. Of the 560 "sail" comprising the navy of 1836, ninety-five were "ships of the line." The largest of these were styled "first-rate ships;" all of them wooden three-deckers, carrying 100 guns each, or more. One of the most difficult problems the Admiralty of that time had to solve was how to ensure a sufficient supply of oak timber for ship-building purposes. Forty full-grown trees to an acre of ground was accounted a good average; at that rate it required the growth of fifty acres to produce enough timber to build one seventy-four-gun ship; and as the oak required at least a hundred years to reach maturity, and the average life of a ship was not much over twenty-five years, the acreage required to produce the entire quantity was enormous. But the prospect of an oak famine was speedily dispelled by the substitution of iron and steel for wood in naval architecture.

Of the 689 vessels of all kinds constituting the

"DUKE OF WELLINGTON" BATTLE-SHIP, 1850.

British navy in 1897, there are only about twenty-two wooden ones, and these are nearly all used either as store ships or training ships, seldom, if ever, to leave their anchorage. And so entirely has the paddle-wheel been superseded by the screw-propeller, there are not left a dozen paddle-steamers in the entire fleet, including the Queen's yachts and a few light-draught river boats. As already mentioned the compound engine was introduced into the navy in 1863. The twin screw was first applied to the *Penelope* in 1868, and has since become universal in vessels of war, the result of these improvements being a marvellous increase of power and speed, with a great saving of fuel. Roughly speaking, a pound of coal is to-day made to produce four or five times the amount of power that it did in 1837.

Experiments had been made with steam power in the navy as early as 1841. In 1845 as many as nineteen sets of screw engines had been ordered for the Admiralty, but it was not until some years later that it came into general use. About 1851 the *Duke of Wellington*,* the *Duke of Marlborough*, the *Prince of Wales*, etc., all full-rigged ships, each armed with 131 " great guns," were fitted with auxiliary steam-engines of from 450 to 2,500 horse-power. The introduction of iron armour-plating—first practised by the French towards the close of the Crimean war— presaged the beginning of the end of "the wooden walls of Old England," and the disappearance forever of the beautiful white wings that had spread themselves out over every sea.

The *Warrior*, completed in 1861, was built entirely

---

* The *Duke of Wellington* was 240.6 feet long, 60 feet beam, 3,826 tons burthen, and 2,500 horse-power. She was engined by Robert Napier & Sons, Glasgow, with geared engines and wooden cogs, and made 10.2 knots an hour on her trial trip in 1853. The *Rattler*, of 1851, was 179½ feet long, 32¾ feet beam, had geared engines of 436 horse-power, and attained a speed of 10 knots.

of iron, protected at vital points by armour-plating four and a half inches in thickness, which, at the time, was supposed to render her invulnerable. She was the precursor of a class of enormous fighting machines, which, however ungainly in appearance, have increased the sea-power of Britain to an incalculable extent. But, alas, for the four and a half inches of armour-plating! Developments in gunnery called for increased thickness of protective armour.

TORPEDO DESTROYER "HORNET," 1896.

The rivalry betwixt gun and armour-plate, keenly contested for years, has not yet been definitely settled; but when ships' guns are actually in use weighing 110 tons and over, capable of throwing a shot of 1,800 lbs. with crushing effect a distance of twelve miles, and, on the other hand, when ships are to be found carrying twenty-four inches of protective iron and steel plating, it seems as if the climax had been nearly reached. In the meantime the insignificant-looking "torpedo destroyer" is coming to the front as one of the most formidable instruments of marine warfare. Although only about 200 feet long, with a displacement of perhaps 250 tons, they have yet a motive power of 5,000 to 6,000 horse-power, and a speed of from 25 to 35 knots an hour. Some of these destroyers are supposed to be strong enough to deal a death-blow to a first-class battle-ship, and all of them are swift enough to overhaul the fastest cruiser on the ocean. The estimation in which they are held by the Admiralty is apparent from the fact that already upwards of one hundred of them are in commission, and many more are being built. Twenty-five destroyers, it is said, can be built for the cost price of one battle-ship, and in actual warfare there would be exposed the same number of lives in fifteen destroyers as in one battle-ship.

Although no great naval battles have taken place to test the power of the steam navy of Britain, it has been occasionally demonstrated in the form of object lessons. The great Jubilee review of 1887 was a magnificent spectacle, when there were assembled at Spithead 135 ships of war, fully armed and manned, and ready to assert Britain's sovereignty on the high seas. Two years later the exhibition was repeated in the presence of admiring Royalty. In January, 1896, shortly after President Cleveland's threatening message to Congress, and while strained relations with Germany had arisen out of complications in South Africa, in an incredibly short space of time the famous "flying squadron" was mobilized and made ready for sea and any emergency that might transpire, without at all encroaching on the strength of the ordinary Channel fleet. The recent naval review in connection with Her Majesty's Diamond Jubilee, however, surpassed any previous display of the kind, not alone as a spectacular event, but as a telling demonstration of sea-power, such as no other nation possesses. On this occasion 166 British steamships of war were ranged in line extending to thirty miles in length, and this without withdrawing a single ship from a foreign station; the only regret expressed on this occasion being that not one of the old "wooden walls" was there with towering masts and billowy clouds of canvas to bring to mind the days and deeds of yore, and to emphasize the remarkable changes introduced by steam.

The following table published by the London *Graphic* exhibits in convenient form the numerical strength of the British navy at the beginning of 1897:

Classification.	Number.	Tons.	Horse-Power.	Officers and Men.	Guns.
Battle-ships, 1st class	29	377,176	355,000	19,291	1,301
" 2nd class	12	114,030	75,000	5,672	346
" 3rd class	11	77,820	57,600	5,487	365
" armoured	18	136,960	116,000	10,386	604
Coast Defence, Iron-clads	16	61,410	30,460	3,211	209
Total armored	86	767,396	634,060	44,047	2,825
Cruisers, 1st class	17	157,950	278,000	10,514	688
" 2nd class	57	243,820	461,100	19,346	1,359
" 3rd class	52	110,685	220,340	10,994	927
Gunboats, Catchers	33	25,940	113,300	2,935	203
" Coast Defence	42	11,828	5,860	1,527	106
Sloops	22	23,305	28,000	2,764	318
Gunboats, 1st class (police)	20	15,810	23,400	1,670	202
Miscellaneous Vessels	24	112,712	202,300	4,998	318
Torpedo Boats and Destroyers	250	25,000	300,000	5,860	690
Grand Total	689	1,494,440	2,266,360	104,855	7,638

First-class battle-ships are vessels of from 10,000 to 15,000 tons displacement, with steam-engines of 10,000 to 12,000 horse-power and attaining a speed of from seventeen to eighteen knots. To this belong the *Magnificent*, the *Majestic*, the *Renown*, the *Benbow*, etc. The first three carry each four 12-inch guns, twelve 6-inch, sixteen 12-pounders, twelve 3-pounders, eight machine guns, and five torpedo tubes. The *Benbow* carries two 16.25-inch guns, each weighing 110 tons, in addition to her armament of smaller pieces. Second-class battle-ships, such as the *Edinburgh* and *Colossus*, are under 10,000 tons, and with 5,500 horse-power develop a speed of about fourteen knots. Third-class battle-ships are represented by the *Hero* and *Bellerophon*, vessels of 6,200 and 7,550 tons respectively.

First-class cruisers include such well-known ships as the *Blake* and the *Blenheim*, each about 9,000 tons with 20,000 horse-power and twenty-two knots speed. The *Powerful* and *Terrible*, also belonging to this class, are among the finest ships in the navy, each 14,200 tons, 25,000 horse-power, twenty-two knots speed, and having crews of 894 men. Additions to the British navy are not made arbitrarily, but with due regard to the enlarged and improved naval armaments of other countries, and with the determination to keep well ahead of all foreign rivals. Accordingly we find that an order was given by the Admiralty in 1897 for the construction of four additional battle-ships and four large cruisers of great speed, the former to be of the *Majestic* type, but with heavier

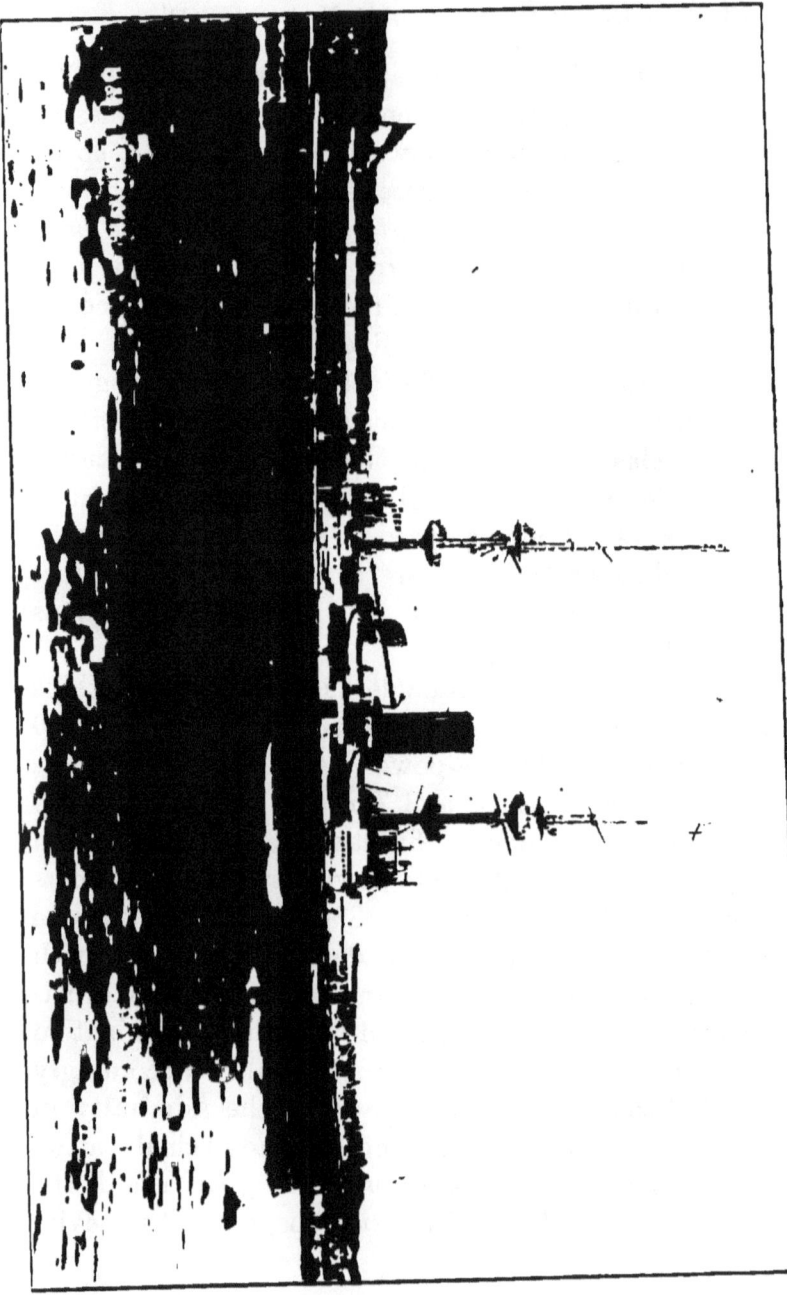

guns, more efficient armour and higher speed, at the same time of slightly less draft, so that if necessary they can pass through the Suez Canal. The cost of a first-class battle-ship, including armament, is about £700,000 sterling or about $3,500,000. A first-class cruiser of the ordinary type costs £450,000, but the *Powerful* and *Terrible*, when ready for sea, are said to have cost £740,000 each. The latest type of torpedo destroyer costs £60,000. The largest projectiles used in the service (as in the *Benbow*) are $16\frac{1}{4}$ inches diameter, weigh 1,820 lbs., and are fired with a charge of 960 lbs. of powder. The average annual expenditure for construction and repairs is between four and five millions, but in 1896 it reached £7,500,000 sterling.

An interesting feature of the Diamond Jubilee review at Spithead, as on former occasions, was the presence of representatives of the mercantile marine in the garb of armed cruisers. By arrangements between the Admiralty and the Cunard, the P. & O., the White Star, and the Canadian Pacific Steamship companies, £48,620 were paid last year in the form of subventions, the vessels so held at the disposal of the Government being the *Campania, Lucania, Teutonic, Majestic, Himalaya, Australia, Victoria, Arcadia, Empress of India, Empress of Japan,* and *Empress of China.*

Many other mercantile steamers besides these are also at the disposal of the Government, being subsidized, and the facilities for converting them into armed cruisers at short notice are most complete, a reserve stock of breech-loading and machine guns being kept in readiness at convenient stations where the transformation can be effected in a few hours. The armament of the *Teutonic* when she appeared at Her Majesty's Diamond Jubilee review consisted of eight 4.7-inch quick-firing guns, and eight Nordenfeldt guns. As an example of how quickly a large auxiliary fleet might at any time be equipped, the case of the *Teutonic* is in point. Leaving New York

"TEUTONIC," ARMED CRUISER, IN 1897.

on Monday, June 14th, with her usual mails and passengers, she reached Liverpool on the 21st. Between that and the 24th she discharged her cargo, was thoroughly cleaned, took on her armour and full complement of naval officers and men, and having on board a host of distinguished guests, was at her appointed place in the review on Saturday, the 26th. Returning to Liverpool, she laid aside her guns, and on the 30th sailed for New York, as if nothing had happened. The *Campania*, which left New York two days later than the *Teutonic*, also appeared at the review in holiday dress, her only armament, however, on this occasion consisting of a large detachment of members of the Houses of Lords and Commons, among whom doubtless were many " great guns."

Marine Distances.

A nautical mile, or "knot," is about 6,082.66 feet; a statute, or land mile, 5,280 feet; the knot is, therefore, equal to 1.1515 mile. The circumference of the earth being divided geographically into 360 degrees, and each degree into 60 nautical miles, the circumference measures 21,600 knots, equal to about 25,000 statute miles. Knots can be readily reduced to statute miles by means of the following table:

Knots	1	2	3	4	5	10	25	100
Miles	1.151	2.303	3.454	4.606	5.757	11.515	28.787	115.148

When the *Lucania* averaged 22 knots, she was running at the rate of $25\frac{1}{3}$ statute miles an hour; her longest day's run (560 knots) was equal to $644\frac{3}{4}$ miles, about the distance covered by an ordinary fast express train on the Canadian Pacific Railway.

The old-fashioned ship's "log" is a piece of wood in the form of a quadrant, loaded with lead at the circumference, to which is attached a line of 120 fathoms or more. Allowance being made for "stray line," the balance is divided into equal distances by knots and small bits of coloured cloth. The distance between each knot is the same part of a mile that 30 seconds is of an hour (the 120th); the length between knots should thus be a trifle over 50 feet. The number of knots run out in half a minute (as measured by the sand-glass) indicate the number of nautical miles the ship is running per hour.

Even express steamships do not always sail between given points exactly as the crow flies. Various reasons lead to the selection of different routes, and even when following the same route, the actual distance run varies a little on each voyage. The Cunard Line, as a precautionary measure, has four sharply defined " tracks " across the Atlantic—two for the westward and two for the eastward voyages—one pair being used in summer and the other in winter, or the ice season.* The northern route, used from July 15th to January 14th, is considerably shorter than the southern route, which is followed from January 15th to July 14th. The distances by these routes are given by the company as follows :

Queenstown to Sandy Hook, by northern track	2,782 knots.
" " " " southern "	2,861 "
Sandy Hook to Queenstown, " northern "	2,809 "
" " " " southern "	2,896 "

Daunt's Rock, Queenstown, being about 244 knots from Liverpool, and Sandy Hook lightship 26 knots from New York, the distance from Liverpool landing-stage to the dock in New York by the Cunard's

*See also p. 90.

MARINE DISTANCES.

northern track is about 3,052 knots, and by the southern track, 3,131 knots; from New York to Liverpool, 3,079 and 3,166 knots, respectively. Captain W. H. Smith says that the shortest distance that can be made between Liverpool and New York is 3,034 knots.

TABLE OF DISTANCES.*

From	To	Distance
Sandy Hook to	Antwerp	3,336 knots.
"	Bremen	3,484 "
"	Copenhagen	3,800 "
"	Genoa	4,050 "
"	Gibraltar	3,200 "
"	Glasgow, *via* North of Ireland	2,941 "
"	Hamburg	3,510 "
"	Havre	3,094 "
"	London	3,222 "
"	Naples	4,140 "
"	Southampton	3,100 "
"	Queenstown	2,809 "
"	Liverpool, *via* northern route	3,088 "
Quebec to Montreal, by the river		160 miles.
" " by the Canadian Pacific Railroad		172 "
"	Rimouski	180 "
"	Belle Isle	747 "
"	St. John's, Newfoundland	896 "
"	Moville, *via* Belle Isle and North of Ireland	2,460 knots.
"	Liverpool, " " "	2,633 "
"	" " Cape Race "	2,801 "
"	" " " and South "	2,826 "
"	Glasgow " Belle Isle and North "	2,564 "
"	" " Cape Race "	2,732 "
"	Queenstown, *via* Belle Isle	2,473 "
Moville to Liverpool		190 "
Halifax to New York		538 "
"	Quebec	680 "
"	St. John's, Newfoundland	520 "
"	Liverpool, *via* North of Ireland	2,450 "
"	" " South "	2,475 "
"	London	2,723 "
"	Glasgow	2,381 "

* Based on a compilation by Captain W. H. Smith.

Halifax to St. John, N.B.	277	knots.
" Portland, Me.	336	"
" Sable Island	169	"
" Boston, Mass	420	"
St. John's, Newfoundland, to Galway, Ireland, which is the shortest land-to-land voyage	1,655	"
Liverpool to St. John, N.B., via North of Ireland	2,700	"
" Portland, Me., " " "	2,765	"
" Boston, Mass., " " "	2,807	"
" Queenstown	244	"
Montreal to Halifax, via Intercolonial Railroad	845	miles.
" " " Canadian Pacific Railroad	756	"
" Boston, " Central Vermont Railroad	334	"
" Portland, Me., via Grand Trunk Railroad	297	"
" New York, via Central Vermont Railroad	403	"
" Toronto, " Grand Trunk Railroad	333	"
" " " Canadian Pacific Railroad	338	"
" " by water	376	"
" Winnipeg, Man, via Canadian Pacific Railroad	1,424	"
" Vancouver, B.C., " " " "	2,906	"
Vancouver to Yokohama, Japan	4,283	knots.
" Shanghai, China	5,330	"
" Hong Kong "	5,936	"
" Honolulu, Hawaii	2,410	"
" Sydney, N.S.W.	6,824	"
Loch Ryan to Quebec, via Belle Isle	2,513	"
" North Sydney, C.B.	2,161	"
" Halifax, N.S.	2,330	"
" St. John, N.B.	2,580	"
Milford Haven to Quebec, via Belle Isle	2,587	"
" Halifax	2,353	"
" North Sydney, C.B.	2,186	"

Sunday at Sea.

As far as circumstances permit, Sunday is observed with as much decorum on shipboard as it is on shore; that is, on the British and American lines. As for the continental steamers, the traveller may expect to become acquainted with a continental Sabbath, which, in most cases, means the ignoring of the day of rest altogether. On our Canadian steamships, weather permitting, public worship is usually held in the saloon, at 10.30 a.m. Sometimes there is an evening service as well, but more frequently an impromptu service of song, much enjoyed by the musical portion of the company, and that is frequently a large proportion of the passengers—ladies especially. The order of service is entirely at the discretion of the captain. In the absence of a clergyman, the captain reads the morning service and the Scripture lessons for the day from the Book of Common Prayer. If there is a Protestant minister on board it is customary to invite him to take the whole service; if there be more than one minister available, each of them may be asked to take part in the service. On the New York liners, as a rule, there is no sermonizing, no matter how many ministers may be on board. The captain and purser read the morning service, or portions of it; a couple of hymns are sung; a collection is taken up for the benefit of the Seamen's Home, or kindred object, and that is all. There are, however, exceptions to this rule. When the captain is prevented by his duties on deck from conducting the service, a clergyman, if there be one among the passengers, is usually asked to assist. A deviation from the rule is often made when a minister of outstanding celebrity happens to be on board. Ministers like the late Dr. Norman Macleod, or Dr. William M. Taylor, would invariably be asked to preach, no matter what line they travelled by. The service-book of the Cunard Company consists of selections from the Book of Common Prayer, with the addition of a form of prayer prepared by the General Assembly of the Church of Scotland, for the use of sailors and persons at sea. A singularly beautiful prayer it is:

"Almighty God, who art the confidence of all the ends of the earth, and of them that are afar off upon the sea; under whose protection we are alike secure in every place, and without whose providence we can nowhere be in safety; look down in mercy on us, thine unworthy servants, who are called to see thy wonders on the deep, and to perform the duties of our vocation in the great waters. Let thine everlasting arm be underneath and round about us. Preserve us in all dangers; support us in all trials; conduct us speedily and safely on our voyage, and bring us in peace and comfort to our desired haven.

"Be pleased to watch over the members of our families, and all the beloved friends whom we have left behind. Relieve our minds from all anxiety on their account by the blessed persuasion that thou carest for them. Above all, grant that our souls may be defended from whatsoever evils or perils may encompass them; and that, abiding steadfast in the faith, we may be enabled so to pass through the waves and storms of this uncertain world, that finally we may come to the land of everlasting rest, through Jesus Christ our Lord. Amen."

The service-book also contains the Psalms of David in prose, and a collection of 107 hymns, including four of the Scotch paraphrases. The hymn most frequently sung at sea is the one beginning with "Eternal Father, strong to save," and next to it, "O God, our help in ages past." Evangelistic services of a less stately kind than in the saloon are often held in the afternoon in the second cabin or steerage, and are usually much appreciated; while in the evening the deck hands will join with groups of emigrants in singing Moody and Sankey hymns, such as "Revive us Again," "Rescue the Perishing," "Whiter than Snow," etc. It is often remarkable to notice how familiar people of diverse creeds and nationalities are with these hymns, and how heartily they unite in singing them.

A favourite text with preachers on shipboard is Rev. xxi. 1: "And there was no more sea." The theme, associated, as it is, with so many fathoms of profundity, has yielded to many forms of treatment. I remember that a young minister, my room-mate, by the way, on his first voyage out from Quebec, chose this for his text, and that he launched out, as well he might, on the charms of the sea in poetical flights of fancy. But the while we were sailing in smooth water. When outside the Straits he laid his head on the pillow and underwent a change of environment, recovering from which, after many days, he vowed that should he ever preach from that text again, he would have something more to say about it. I remember, too, that an elderly gentleman—a Presbyterian of the Presbyterians—was asked by the captain to preach one Sunday morning. He readily complied, taking it for granted that he was to conduct the whole service. Imagine his chagrin when an Anglican brother unexpectedly appeared on the scene and went through the whole of the long service of the Church of England. With the utmost composure, Πρεσβύτερος simply ignored the beautiful liturgical service, commenced *de novo*, and went through the whole service afresh, in orthodox Presbyterian fashion, to the surprise of the congregation and the discomfiture of the waiters, whose time for setting the lunch-table was long past.

A distinctive and pleasing feature of these Sunday services at sea, especially in the larger steamships, which often carry more passengers than would fill an ordinary church, is the heartiness with which the representatives of various religious denominations unite in the services. The lines of demarcation that separate them when ashore seem to be lost sight of at sea. Casual acquaintanceship here frequently ripens into closer friendship; people begin to see eye to eye, and soon the conviction grows stronger that the doctrinal points on which all professing Christians are agreed are much more important than the things about which they differ. It would do some narrow-minded souls a world of good to spend a few Sundays at sea.

The office for the burial of the dead at sea is very solemn and affecting. In the days of sailing ships, when voyages lasted so much longer, deaths from natural causes at sea were more frequent than now. But the order of service is the same. The body of the deceased person might be sewed up in a hammock —indeed, it usually was—or the carpenter may have made a rough coffin for it. In either case it was heavily loaded with iron at the foot. A stout plank with one end resting on the bulwark forms the bier on which is laid the corpse, covered with an ensign. The captain, the chief engineer, the ship's doctor and purser, with a detachment of the crew, and a few of the passengers, make up the funeral party. Portions of the Church of England's beautiful service for the burial of the dead are read: "I am the Resurrection and the life." . . . "I know that my Redeemer liveth." . . . "We brought nothing into this world and it is certain we can carry nothing out." . . . "Man that is born of a woman hath but a short time to live," etc. The ship's engines are then stopped for a few seconds while the service proceeds —" We therefore commit his body to the deep, looking for the resurrection of the body when the sea shall give up her dead."

The ensign is removed. The inward end of the plank is raised, and the mortal remains are plunged into the greatest of all cemeteries; sometimes with scant ceremony, perhaps, but always impressing on the mind of the spectator a deeply pathetic incident that will never be forgotten.

> "And the stately ships go on
> To their haven under the hill;
> But O for the touch of a vanished hand,
> And the sound of a voice that is still."

Icebergs and Tidal Waves.

Icebergs and bewildering fogs, as has been already said, are a large element of danger in the St. Lawrence route. The passengers who sailed with me on the

Lake Superior, from Montreal on July 1st, 1896, will not soon forget the magnificent display of icebergs which they witnessed on the Sunday following. From early morning until midnight, for a distance of more than 250 miles, the ship's course lay through an uninterrupted succession of icebergs—a procession, it

H. M. YACHT "VICTORIA AND ALBERT," 1855.
2,470 tons ; 2,980 h. p.: speed, 16.8 knots ; armament, 2 six-pounders ; crew, 151 men.

might be called, on a grand scale of masses of ice in all manner of fantastic shapes and of dazzling whiteness—travelling to their watery graves in the great Gulf Stream of the south. Mountains of ice, some of them might be called. On one of them a grisly bear was alleged to have been seen sulkily moving to and fro, as if meditating how, when and where his romantic voyage was to come to an end. The day was calm and cloudless—a perfect day for such a marvellous exhibition. It might have been otherwise, and how different may be imagined from reading what appeared in the English papers a few weeks later—the account of a ship's narrow escape from destruction in this identical locality:

"STRUCK AN ICEBERG.—The SS. *Etolia* on her voyage from Montreal to Bristol narrowly escaped destruction from collision with an iceberg twenty-four hours after leaving the eastern end of Belle Isle straits. A dense fog had set in, the lookout was doubled, and the engines slowed; presently the fog lifted, but only to come down again thicker than ever. In a very short time the lookout called out, 'Ice ahead!' The engines were promptly stopped, then reversed at full speed. Meanwhile the towering monster bears down on the ship and in a few seconds is on top of it. It was a huge berg, rising high above the masts of the steamer, which it struck with such a crash that some three hundred tons of ice in huge pieces came down on the forecastle. Fortunately most of it rebounded into the sea, but some forty or fifty tons remained on the ship's deck. The ship trembled under the blow from stem to stern; her bows were smashed in, but the leakage was confined to the fore-peak. In this battered condition the *Etolia* lay without a movement of the engines for thirty-six hours until the fog cleared, when Captain Evans had the satisfaction of proceeding on his course and bringing his passengers and crew safely into Bristol harbour."

A still more serious disaster was reported on August 25th of the same year (1896):

"The captain of the steamer *Circassia*, of the Anchor Line, had a story to tell, on her arrival at quarantine early this morning, of picking up a captain and his twenty-two men on the high seas from three open boats. It was Captain Burnside and the entire crew of the British tramp steamer *Moldavia*, bound from Cardiff to Halifax with coal, who were rescued by the timely approach of the *Circassia*. During the dense fog over the sea on last Wednesday, the *Moldavia* ran into a huge iceberg and stove her bows so badly that she began to fill rapidly. It was 5.30 o'clock in the afternoon. As soon as a hasty examination showed that it would be impossible to save his ship, Captain Burnside ordered the lifeboats provisioned and cleared away, and as soon as it could be done the steamer was abandoned and shortly afterwards sank. The lifeboats kept together and watched for a passing vessel, and thirty-five hours later the *Circassia's* lights were seen approaching. Blue lights were at once shown by the occupants of the lifeboats, and the *Circassia* altered her course. When near enough, Captain Boothby, of the *Circassia*, hailed the lifeboats and told the men that he would pick up the boats and their occupants. Accordingly the davits' tackle were lowered, and as each lifeboat approached she was hooked on and raised bodily, occupants and all, to the deck of the *Circassia*."

The icebergs of the North Atlantic are natives of Greenland or other Arctic regions where glaciers abound. They carry with them evidence of their terrestrial birth in the rocks and debris with which they are frequently ballasted. The glacier, slowly moving over the beds of rivers and ravines, ultimately reaches the seaboard, to be gradually undermined by the action of the waves, and, finally, to fall over into deep water and be carried by winds and currents into the open ocean. In their earlier stages icebergs are constantly being augmented in size by storms of snow and rain, and by the freezing of the water washed over them by the waves. They are of all sizes, from a mere hummock to vast piles of ice half a mile in diameter, and showing an altitude above the sea of two or three hundred feet, sometimes rising to a height of five and even six hundred feet, and that is scarcely more than one-eighth of the whole mass, for a comparatively small portion only of the bulk projects above the surface, as may be plainly seen by dropping a piece of ice in a tumbler full of water. In proof of this, it is by no means uncommon to find icebergs of ordinary dimensions stranded in the straits of Belle Isle in seventy or eighty fathoms of water. Being frequently accompanied by fog—of which they may be the chief cause—they are often met with unawares, though their nearer approach is usually discovered by the effect which they produce on the air and the water surrounding them, suggesting to the careful navigator the frequent use of the thermometer to test the temperature of the water where ice is likely to be encountered. They are seldom met with below the 40th parallel.

Field-ice, covering a surface of many square miles, with a thickness of from ten to twenty feet, is frequently fallen in with off the coasts of Labrador and Newfoundland. Though less dangerous to navigation than the iceberg, it is often a serious obstruction. Vessels that incautiously run into a pack of ice of this kind, or have drifted into it, have often found themselves in a *maze*, and have been detained for weeks at a time, and not without some risk to their safety in heavy weather.

TIDAL WAVES.—Notwithstanding elaborate treatment of the subject by hydrographers, stories about ocean tidal waves are most frequently relegated by landsmen into the same category with tales of the great sea-serpent. Sailors, however, have no manner of doubt as to their existence and their force. During violent storms it has been noticed that ocean waves of more than average height succeed each other at intervals—some allege that every seventh wave towers above the rest. Be that as it may, there is no doubt that a sudden change of wind when the sea is strongly agitated frequently produces a wave of surpassing magnitude. Other causes, not so obvious, may bring about the same result, producing what in common parlance is called a "tidal wave." This is quite different from the tidal wave proper, which periodically rushes up the estuaries of rivers like the Severn, the Solway, the Garonne, the Hoogly and the Amazon. In the upper inlets of the Bay of Fundy, where the spring-tides rise as high as seventy feet, the incoming tide rushes up over naked sands in the form of a perpendicular white-crested wave with great velocity. The tidal wave of the Severn comes up from the Bristol Channel in a "bore" nine feet high and with the speed of a race-horse, while the great bore of the Tsien-Tang-Kiang in China is said to advance up that river like a wall of water thirty feet in height, at the rate of twenty-five miles an hour, sweeping all before it.* The ocean tidal wave dwarfs these and all other waves by its huge size and tremendous energy. The effective pressure of such a wave being estimated at 6,000 pounds to the square foot, it is easy to understand how completely it becomes master of the situation when it topples over on the deck of a ship. Only once in the course of a good many voyages has the writer been an eye-witness of its tremendous force. The occasion was thus noticed in the New York papers of the 2nd and 3rd of August, 1896:

"The American liner *Paris* and the Cunarder *Etruria*, which arrived on Saturday, had a rough-and-tumble battle before daylight on Tuesday morning with a summer gale that had an autumn chill and a winter force in it. The wind blew a whole gale and combed the seas as high as they are usually seen in the cyclonic season. The crest of a huge wave tumbled over the port bow of the *Etruria* with a crash that shook the ship from stem to stern, and momentarily checked her speed; a rent was made in the forward hatch through which the water poured into the hold, flooding the lower tier of staterooms ankle-deep. The ship's bell was unshipped, and it carried away the iron railing in front of it, snapping iron stanchions two inches in diameter as if they had been pipe-stems. The *Paris*, about the same hour and in the same locality, shipped just such a sea as that which hit the *Etruria*, but received less damage. It fared much worse, however, with the sailing ship

* "Encyclopedia Brit.," Vol. xvii., p. 581, 8th Ed.

Ernest, from Havre, which was fallen in with on the morning of the gale showing signals of distress. The French liner *La Bourgogne*, came to her rescue and gallantly took off the captain and his crew of eleven men, abandoning the shattered ship to her fate with ten feet of water in her hold."

It is not often that a tidal wave visits the St. Lawrence, but in October, 1896, the SS. *Durham City*, of the Furness Line, when off Anticosti, was struck by a big wave which carried away her deckload, including sixty-eight head of cattle and everything movable. It was only one sea that did the damage, but it made a clean sweep.

By a figure of speech, ocean waves are frequently spoken of as running "mountains high," and the popular tendency is doubtless towards exaggeration. The estimate of experts is that storm waves frequently rise to forty feet, and sometimes even to sixty or seventy feet in height from the wave's base to crest.

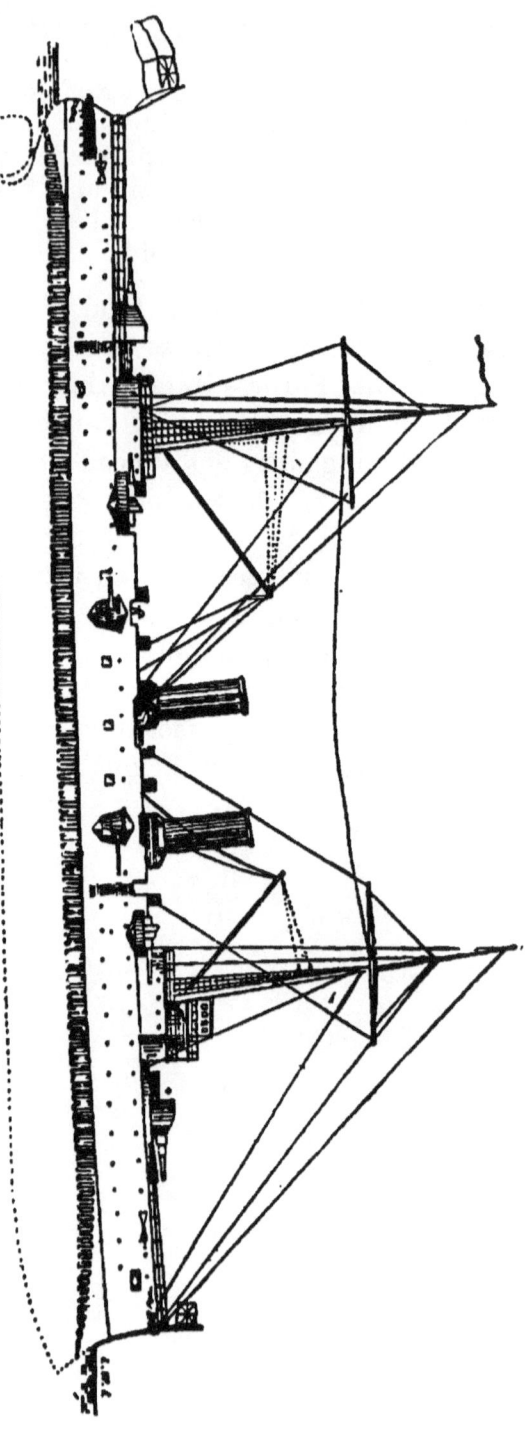

H. M. SS. "CRESCENT."

Presented by publishers of the "Star Almanac," Montreal, 1896.

This outline represents one of the smaller types of British warships, known as first-class cruisers. The *Crescent* was launched at Portsmouth in 1892, and cost £383,068. She is 300 feet long and 60 feet beam. Her tonnage is 7,700 tons; her indicated horse-power 12,000, and her speed 19.7 knots an hour. Her armament consists of one 22-ton gun, twelve 6-inch quick-firing, twelve 6-pounder do., five 3-pounder do., seven machine guns and two light guns. The *Crescent* was for several years the flagship of Vice-Admiral James Elphinstone Erskine, on the North American and West Indies Station, and is consequently well known in Canadian waters. She visited Quebec several times.

CHAPTER VII.

THE ST. LAWRENCE ROUTE.

The Allan, Dominion, Beaver, and other Canadian Lines of Ocean Steamships—Sir Hugh Allan—A Fast Line Service, etc., etc.

WERE it not that the St. Lawrence is hermetically sealed for five months of the year, it would undoubtedly be a more formidable rival to the Hudson than it now is. That great drawback, however, is not the only one. The navigation of the St. Lawrence has always been somewhat difficult and hazardous. The seven hundred and fifty miles of land-locked water from Quebec to Belle Isle is notorious for swift and uncertain tides and currents, for treacherous submerged reefs and rocks, and shoals in long stretches of the river, for blinding snow-storms and fields of floating ice in the lower reaches at certain seasons of the year, for icebergs which abound on the coasts of Labrador and Newfoundland, and for bewildering fogs. With such a combination of difficulties it is not to be wondered at that shipwrecks have been frequent; that they have not been more numerous must be mainly attributed to good seamanship and an intimate knowledge of the route. Nautical appliances and charts are very much better than they were thirty or forty years ago. The efficiency of the lighthouse system has been greatly increased, and, what is vastly important, the masters of mail steamers are no longer restricted to time, but on the contrary are instructed that whenever the risk of life or of the ship is involved, speed must be sacrificed to safety.

The St. Lawrence route has some advantages over the other. It is nearly five hundred miles shorter from Quebec to Liverpool than from New York. Other things being equal, passengers by this route have the advantage of 750 miles of smooth water at the beginning or end of their voyage, as the case may be. For these and other reasons many prefer the St. Lawrence route. It has become popular even with a good many Americans, especially from the Western States, and will certainly become more so if the contemplated "fast service" is realized, by which the ocean voyage—from land to land—would be curtailed to three days and a half !

In the discussions that have arisen on the subject, the danger of running fast steamers on this route has, in many instances, been unduly magnified. Past experience tends to show that the actual risk is not necessarily increased by fast steaming. Shipwrecks in the Gulf of St. Lawrence during later years have been confined to cargo and cattle steamers. Not one of the faster mail boats has been lost during the last sixteen years. The chief difficulty in the way of establishing a twenty-knot service for the St. Lawrence is that of the ways and means. Would it pay ? Certainly not by private enterprise alone, but the favour with which the project is regarded by the

Imperial and Dominion Governments leaves little doubt that it will be accomplished in the near future.

Captain W. H. Smith, formerly Commodore of the Allan Line, in command of the *Parisian*, and who,

CAPTAIN W. H. SMITH, R.N.R.

from long service on this route, is well qualified to express an opinion, states in his report to the Government that he sees no reason why there should not be a fast line of steamers to the St. Lawrence. "If," he says, "the St. Lawrence route is selected for the proposed fast line, there should be no racing in competition with other large steamers, and the same amount of caution must be taken which has been exercised of late years by senior officers of the Allan and other lines trading to Canada; and it will be absolutely necessary for the safety of navigation that the commanders and officers of any new company should be selected from the most experienced officers of existing lines."

In 1853 a Liverpool firm, Messrs. McKean, McLarty and Lamont, contracted with the Canadian Government to run a line of screw steamers, to carry Her Majesty's mails, twice a month to Quebec in summer, and once a month to Portland during the winter, for which the company was to receive £1,238 currency per trip, under certain conditions, one of which was that the ships should average not more than fourteen days on the outward, nor more than thirteen days on the voyage eastward. The ships of the first year were the *Genova*, 350 tons; *Lady Eglinton*, 335 tons; and *Sarah Sands*, 931 tons. Their average passages were wide of the mark. Next year the *Cleopatra*, *Ottawa* and *Charity* were added to the line. The *Cleopatra* made her first trip to Quebec in *forty-three days*; the *Ottawa* never reached Quebec at all, but after dodging about some time among the ice at the mouth of the St. Lawrence, made for Portland. The *Charity* reached Quebec in twenty-seven days. As a matter of course the contract was cancelled.

The Allan Line.

The failure of the Liverpool firm to fulfil their contract opened the way for Canadian enterprise, and the man who was destined to see it carried out to a successful issue was already awaiting his opportunity. That man was Hugh Allan (the late Sir Hugh), a man of intense energy and force of character. The Allans came honestly by their liking for the sea and ships. Their father, Alexander, was a ship-owner, and himself the well-known captain of the *Favourite*, one of the most popular vessels then sailing from the Clyde to the St. Lawrence. The five sons were born at Saltcoats, in sight of the sea. Two of them, James and Bryce, followed the sea for a number of years and reached the top of their profession. Alexander took up the shipping business established by his father in Glasgow, where he was afterwards joined by his eldest brother, James, under the firm name of James and Alexander Allan. Bryce, on retiring from the sea, became head of the shipping house in Liverpool. Hugh, the second son, became a partner in the well-known firm of Miller, Edmonstone & Co., afterwards changed to Edmonstone, Allan & Co., Montreal. His brother Andrew joined the firm some years later, when its name was changed to that of Hugh and Andrew Allan. The three firms, in Glasgow, Liverpool and Montreal, had become the owners and agents of a large fleet of sailing ships; but the time came when it was evident that mails and passengers must be carried to Canada, as elsewhere, by steam power.

The opening of the St. Lawrence and Atlantic Railway between Montreal and Portland in 1852 was one of the most important events in the commercial history of Canada. It gave Montreal a

CAPTAIN McMASTER.

winter port; for as yet neither Halifax nor St. John had any railway communication with the western provinces. Given a good winter port, there seemed to be no reason why a line of steamships should not be established to ply between Liverpool and Montreal in summer, with Portland for the winter terminus. The Allans, seeing that the time had come for a new departure, succeeded in forming a joint stock company, under the name of the Montreal Ocean Steamship Company. As its name implied, it was virtually a Canadian enterprise. The principal shareholders, besides the Allans, were Messrs. William Dow, John G. Mackenzie and Robert Anderson, of Montreal; George Burns Symes, of Quebec, and John Watkins, of Kingston. A few years later the Allans became sole owners of the concern, which then became known as the ALLAN LINE.

The first two steamers of the Montreal Ocean Steamship Company were the *Canadian* and *Indian*, built by the famous Dennys, of Dumbarton. They were pretty little iron screw steamers, of about 270 feet in length, 34 feet wide, and of 1,700 tons burthen each. The *Canadian* made her first voyage to Quebec in September, 1854, but the Crimean war having commenced, steamers of this class were in demand, and these two were taken into the service and profitably employed as government transports as long as the war lasted. In 1874 the *Sarmatian* and the *Manitoban* of this line were similarly employed to convey troops to the west coast of Africa, to take part in the Ashantee campaign. On both occasions they did excellent service.

When the Canadian Government next advertised for tenders for carrying the mails, an agreement was made with the Allans by which they were to receive £25,000 a year for a fortnightly service in summer and a monthly one in winter. Two other boats, similar to the *Canadian* and *Indian*, were built by the Dennys—the *North American* and *Anglo-Saxon*. The new service was commenced in April, 1856, by the SS. *North American*, which arrived in the port of Montreal on the 9th of May. Two years later it was decided to establish a weekly service, the Government promising an increased subsidy of $208,000 per annum. This implied double the number of ships; accordingly, four others were built, the *North Briton, Nova Scotian, Bohemian* and *Hungarian*, all after the same model as the pioneer ships, but 300 feet long and 2,200 tons register. Their speed was from 11 to 13 knots in smooth water, and even in heavy weather they seldom fell short of 8 knots an hour. Their average passages westward from Liverpool to Quebec were 11 days, 5 hours; eastward, 10 days, 10 hours. The quickest passage eastward was made by the *Anglo-Saxon*, in 9 days, 5 hours, and westward, by the *Hungarian*, in 9 days, 14 hours. In the same year (1859) the Cunard Line to Boston averaged 12 days, 19 hours going west, and 10 days, 15 hours eastward. The average speed of the Canadian steamers during the entire season of the St. Lawrence navigation in that year was $9\frac{1}{2}$ knots. At this time there were already twelve different lines of steamships plying across the Atlantic, affording almost daily conmunication between England and America by steam.

In 1859 the company represented that, owing to the depression in trade, they were unable to continue the service, without further assistance. The Canadian Government stood by this Canadian enterprise, and doubled the subsidy in consideration of the increased service, which was admitted on all hands to be a complete success. The new ships were beautiful models and well adapted to the trade; but the company had to learn from bitter experience how hazardous that trade was. To say nothing of minor accidents, up to the year 1885 no less than fourteen of their steamers had come to grief. Since that time, singularly enough, none of this line has been lost, though many belonging to other lines have been wrecked.

The *Canadian*, Captain Ballantine, on her first trip to Quebec, in June, 1857, through the negligence of her pilot, was stranded on South Rock, off the Pillar Lighthouse, forty-five miles below Quebec. No lives were lost, but the ship defied every effort to float her. The *Indian*, Captain Smith, bound for Portland, in December, 1859, struck a rock off Marie Joseph Harbour, seventy-five miles east of Halifax, and went to pieces. Every effort was made to save the lives of the 447 persons that sailed in her, but twenty-three perished. The *Hungarian*, Captain Jones, on the night of February 20th, 1860, during a blinding snow-storm, struck on the South-West Ledge near Cape Sable Island, 130 miles east of Halifax. Every soul on board, to the number of 237, perished with the ship. The cause of this sad disaster is not correctly known. The captain was one of the best seamen in the Allan Line, but it has been stated that the light upon Cape Sable was not exhibited that night, in consequence of the sickness of the light-keeper, who is said to have confessed this on his death-bed.

The second *Canadian*, Captain Graham, came in contact with a piece of submerged ice, outside the Straits of Belle Isle, in July, 1861. The ship was proceeding cautiously, but so hard and sharp was the ice, a rent was made in the ship's side below the water-line, and it was soon seen that she was done for. This is how she went down, as told by Captain Graham: "The wind had increased to a gale. About 9.30 a.m. we came up to heavy field ice closely packed. We had been going half-speed till we saw the ice, when we stopped altogether, then turned her head to the west, steaming slowly through a narrow passage between heavy ice on the starboard side and what appeared to be a light patch of ice on the port side, which scratched along the bow for sixty feet. The concussion was very slight, and I had no apprehension of any damage; went below to see what was wrong, and found the water rushing along the main deck and up the hatch-way. The boats were ordered out, and the ship headed for land full speed. She soon began to settle down forward, taking a list to starboard, when the engines were stopped and the boats lowered. Immediately after leaving her, the ship with a plunge dropped five or six feet by the head, and then directly afterwards her stern flew up in the air, and she went down head foremost." The mail-master, nine of the crew and twenty-six passengers went down with the ship.

The *North Briton*, Captain Grange, was wrecked in November, 1861, on one of the Mingan Islands, north of Anticosti (the usual track for steamers at that time). There was no loss of life. The *Anglo-Saxon*, Captain Burgess, in April, 1863, was stranded in Clam Cove, three miles from Cape Race, during a dense fog. A heavy sea rolling in drove her farther on the rocks, from which she eventually slid off and sank in deep water. The captain, some of the officers, and many of the passengers and crew were carried down into the vortex of the ship, and were drowned to the number of 238 souls.

The *Norwegian*, Captain McMaster, in June, 1863, was totally wrecked on St. Paul's Island, at the entrance of the Gulf. A dense fog was prevailing. The passengers and crew, numbering about 420, were all saved. The *Bohemian*, Captain Borland, struck on the Alden Ledges, off Cape Elizabeth, near Portland, in February, 1864, when twenty passengers were drowned. The *Dacian* was wrecked near Halifax, April 7th, 1872. In the same year the *Germany* went ashore at the mouth of the Garonne River, near Bordeaux, France, and was totally wrecked, with the loss of thirty lives. The *St. George*, Captain Jones, was lost on the Blonde Rock, south of Seal Island, N.S. The *Jura* stranded on Formby Bank, at the entrance to the Mersey, in 1864. The *Moravian*, Captain Archer, was wrecked on Mud Islands, near Yarmouth, N.S., in December,

1881. The *Hanoverian*, Captain Thompson, struck a rock at the entrance of Nepassey Bay, Newfoundland, and was totally lost, but all hands were saved.

The *Pomeranian*, Captain Dalziel, a fine ship of 4,364 tons, in 1893 survived one of the stormiest Atlantic voyages on record. She sailed from Greenock for New York, March 27th. After eight days battling with a furious gale, when about twelve hundred miles west of Ireland, she was well-nigh overwhelmed by a tremendous wave, which made a clean sweep of the deck. The bridge, the charthouse, the saloon, the steam-winch, the ventilators, everything between the foremast and the funnel, were hurled overboard, a mass of wreckage. The captain and a saloon passenger were so severely injured that both died in a few hours. The second and fourth officers, who were on the bridge, were swept into the sea and drowned, as were the rest of the cabin passengers, one intermediate, and four of the crew—twelve persons in all. Three of the lifeboats were carried away and two were smashed, leaving only one available for service. The whole of the nautical instruments, books and charts had gone overboard, the steering gear was badly wrecked, and the only compass left was that in the steering-house aft. The first officer, Mr. McCulloch, on whom the command now devolved, seeing the crippled condition of the ship, turned her head homewards, a thing not easily done in such a sea, and eventually returned to the Clyde in a gale of wind.

It is doubtful if there is another shipping company

in existence that would have withstood the strain put on the Allan Line by such a succession of disasters; but so far as outsiders are aware the Allans never lost courage. They were bound to succeed in the long-run, and they did. When ships could not be built quickly enough to take the places of those that

"THE PARISIAN," 1881.

had been lost at sea, they bought of others ships ready-made, meanwhile resolving to reinforce their fleet with larger and in every way better boats than heretofore. The *Norwegian* and *Hibernian*, of 2,400 tons each, were launched from Denny's yard in 1861. In 1863 Steeles of Greenock built for them the

Peruvian and the *Moravian*, both very fine ships. The *Nestorian* and the *Austrian*, of 2,700 tons each, built by Barclay & Curle, Glasgow, are both good ships now after thirty years' service. The *Sarmatian* and *Polynesian* (now *Laurentian*), about 4,000 tons each, came out in 1871 and 1872, and proved excellent boats. The *Circassian*, 3,724 tons, was launched in 1873, and the *Sardinian* in 1875. The *Parisian*, the finest of the fleet, was built by Robert Napier & Sons, Glasgow, in 1881, and took her place on the line the following year. She is built of steel, the bottom being constructed of an inner and outer skin five feet apart, the space thus enclosed being available for water ballast and also a protection from the perils of collision. The Allans were the first to apply this kind of build to Atlantic steamers, and were also the first to build such steamers of steel. The general dimensions of the *Parisian* are: length over all, 440 feet; breadth, 46 feet; moulded depth, 36 feet; with a gross tonnage of 5,365 tons. Her machinery is capable of developing 6,000 indicated horse-power. Although she has neither twin screws nor triple expansion engines, she has done her work remarkably well, maintaining an average speed of about fourteen knots. Her fastest voyage from Moville to Rimouski was made in 1896, viz., 6 days, 13 hours, 10 minutes, corrected time. Her best day's run on that voyage was 359 knots. Her career has been a remarkable one: in these seventeen years she has not met with an accident, and is consequently a very popular ship. She is fitted for 160 saloon passengers in the most complete and comfortable manner, and there seems to be always room for one more. On a recent occasion the *Parisian* brought over 255 cabin passengers. She can easily accommodate 120 second-class and 1,000 steerage passengers. She carries a large cargo and is a very fine sea boat.

The fleet of the Allan Line consists at present of thirty-four steamers, aggregating 134,937 tons. In addition to the weekly line between Liverpool and Montreal, regular weekly services are maintained from Montreal, and also from New York, to Glasgow; the London, Quebec and Montreal service is fortnightly in summer; there is also a direct service between Glasgow and Boston fortnightly, and regular communication between Liverpool, Glasgow and Philadelphia, as well as with River Plate and other ports.

Some of the freight and cattle-ships of the Allan Line are large and fine vessels, such as the *Buenos Ayrean*, 4,005 tons, built at Dumbarton in 1879—one of the first ships ever constructed of steel. The *Carthaginian* and *Siberian* are both 4,000-ton ships, specially adapted for the cattle trade. The *Mongolian* and *Numidian*, of 4,750 tons each, are model ships in the class to which they belong. A few years ago the Allans acquired the State Line, plying between Glasgow and New York. Two of these, the *State of California* (5,500 tons) and the *State of Nebraska* (4,000 tons), are excellent ships with good accommodation for large numbers of passengers. The two oldest ships of the line in commission are the *Waldensian* (formerly *St. Andrew*), built in 1861, and the *Phœnician* (formerly the *St. David*), built in 1864, both of which are still doing service in the South American trade.

The last of the sailing ships owned by the Allans was wrecked in a dense fog near Astoria, at the mouth of the Columbia River, Washington, U.S., on the 19th of March, 1896. The *Glenmorag* was a fine iron clipper ship of 1,756 tons register, built at Glasgow in 1876, and up to the time of her final disaster had been exceptionally fortunate and successful. Captain Currie, who commanded her, was widely known and has a first-rate reputation as a sailor, but in an evil hour of a dark, dirty night, when making for Portland, Oregon, he was startled by the sudden cry from the man on the look-out, "Breakers on the port bow," and while in the act of wearing the ship around she went broadside on the rocks. Two of the crew were killed and four injured severely while attempting to get ashore.

It has been announced that the Allans have at present under construction on the Clyde four magnificent steel steamships for the Canadian freight and passenger trade. Three of these are vessels of 10,000 tons, and the fourth of 8,800 tons. All of them are to be fitted with triple expansion engines and twin screws. The three larger ones are each over 500 feet in length, with 60 feet breadth of beam, and are designed to develop an average speed of sixteen knots, which means that they are expected to make the voyage from Liverpool to Montreal in about 7¼ days mean time — actually a quicker service for Canada than obtains at present with 20-knot steamers *via* New York. With ample accommodation for a large number of passengers, these ships will have room for 8,000 to 9,000 tons of freight and the most approved appliances for the rapid handling of cargo.

Sir Hugh Allan of Ravenscrag, to whom Canada is chiefly indebted for the magnificent Allan Line of steamships, was born at Saltcoats, Ayrshire, Scotland, September 29th, 1810. He came to Canada in 1826 and entered into business as already stated. His whole life was one of incessant activity. He was founder of the Merchants' Bank of Canada and its president, and the President of the Montreal Telegraph Co., and many other important commercial institutions. Sir Hugh was knighted by Her Majesty the Queen, in person, in July, 1871, in recognition of his valuable services to the commerce of Canada and the Empire. He died in Edinburgh, suddenly, December 9th, 1882, and was buried in Mount Royal cemetery, Montreal. Sir Hugh was a man, very emphatically, *sui generis*. Quick to arrive at his conclusions, he was slow to abandon them; where he planted his foot there he meant it to stay. A keen and enterprising man of business, he accumulated a princely fortune. To those who knew him only on the street or in the Board-room he might, perhaps, seem curt and brusque. His conscious power of influencing others made him almost necessarily dogmatic and dictatorial, but in private life he was one of the most amiable, kind-

Sir Hugh Allan.

hearted and genial of men. He was a staunch Presbyterian, a liberal supporter of the Auld Kirk of Scotland in Canada, and in his younger days devoted much time in promoting its interests.

The brothers Bryce and James died several years before Sir Hugh. Alexander died in Glasgow in 1892. Mr. Andrew Allan, now the senior partner of the Montreal firm, was the youngest of the five brothers, and is the only survivor of them. Mr. Allan was born at Saltcoats, December 1st, 1822, and came out to Canada in 1839. He married a daughter of the late John Smith, of Montreal (a sister of Lady Hugh Allan). Mrs. Allan died in 1881, leaving a large family. Two of the sons, Messrs. Hugh H. and Andrew H., are associated with their father and with Messrs. Hugh Montagu and Bryce J. Allan, sons of the late Sir Hugh, in managing the extensive business of the Canadian branch of the Allan Line. Mr. Allan has filled many of the posts of honour and responsibility formerly occupied by Sir Hugh, and earned for himself the golden opinions of his fellow-citizens.

The first four captains of the Allan Line were Andrew McMaster, of the *Anglo-Saxon*, Thomas Jones, of the *Indian*, William Ballantine, of the *Canadian*, and William Grange, of the *North American*. Captain McMaster was born at Stranraer, Wigtonshire, in 1808. After serving a five years' apprenticeship on board the East Indiaman, *Duke of Lancaster*, at the modest rate of £2 for the first year, and £20 for the full term of his indentures, he got command of the brig *Sir Watkin*, sailing from Islay with 240 of the clan Campbell as passengers. One-half of these were landed at Sydney, Cape Breton, and the other half at Quebec. The hardships of the emigrants in those days were excessive, as they had to provide their own food and bedding, and were allotted places on the stone ballast to do the best they could for themselves. In 1845 Captain McMaster was placed in command of the clipper barque, *Rory O'More*, for which Edmonstone & Allan were the agents. Leaving Montreal in the summer of 1846, owing to the lowness of water the yards and topmasts were sent down and floated alongside, while cables, chains and other rigging were put into lighters to enable the vessel to traverse Lake St. Peter, drawing nine feet of water! His next command was the ship *Montreal* of 464 tons, at that time the largest of the Montreal traders. In 1856 he was placed in command of the first SS. *Canadian*, and successively of each new ship as she was launched. In 1864 he retired from the sea, and entered the shipwright business in Liverpool. He died in the Isle of Man in 1884.

Of the subsequent captains of this line I can only mention the names of those with whom I remember having sailed and made their acquaintance. None of them left a more lasting impression on my memory than John Graham, the genial captain of the second *Canadian*, and of the *Sarmatian* when he retired from the service and the sea in 1885. It was he who so often and so strenuously discussed the desirability of throwing a dam across the Straits of Belle Isle that he actually came to believe in it himself as a possibility in the near future, by which in his estimation the climate of Canada was to be assimilated to that of the south of France. That was his fad. But take him all in all, he was as fine a man as one could

CAPTAIN JOHN GRAHAM.

desire to meet. He was a grand sailor. When his examination before the Nautical Board was concluded *in re* the loss of the *Canadian*, his certificate was handed back to him with the remark, "Sir, you did your duty like a noble British seaman." The dangers incident to a seafaring life never disturbed his equanimity, for he had long been ready to " go aloft " at a moment's notice.

James and Hugh Wylie were both quiet, unassuming men who understood their business thoroughly.

CAPTAIN JAMES WYLIE.

The former rose to be the commodore of the fleet. On retiring from the command of the *Parisian*, the citizens of Montreal honoured him with a banquet and an address, congratulating him on his remarkably successful career. Hugh retired from the command of the *Polynesian* shortly after a serious accident that befell his ship on the river, through the carelessness of his pilot. James was noted for his caution, of which a somewhat humorous illustration was given one dark night when the *Parisian* was speeding down the Gulf of St. Lawrence. Some of us were still pacing the deck, though it was near midnight, when suddenly the engine stopped. To the uninitiated there is nothing more alarming than that; but at this hour most of the passengers were fast asleep. There followed a few minutes of profound silence. The sea, until now as black as ink, had all at once become white and glistering. Had we run into a field of ice? To the captain, who was at his post on the bridge, and to the double look-out on the forecastle it must have had that appearance; but it proved to be only schools of herring or mackerel disporting themselves on the surface of the water, causing a brilliant phosphorescent illumination of the sea. It spread over a large surface and had all the appearance of field ice, precisely where such danger is to be apprehended. The ship sailed on: but none of us dared to ask then, nor next morning, why she had stopped.

Frederick Archer, Lieut. R.N.R., successively in command of the *St. Andrew*, the *Manitoban*, and the *Moravian*, was made of sterner stuff than the average sea-captain. It required more than one voyage to become acquainted with him, but once in his good graces the passenger was all right. He was the strictest disciplinarian of the whole staff. Regularly as on a man-of-war, his sailors marched into the saloon on Sunday mornings in their best rigs to attend divine service. In the absence of a clergyman none could use the Book of Prayer more effectively than Capt. Archer. He died at sea in the prime of life.

William H. Smith, Lieut. R.N.R., son of late Commander John S. Smith, R.N.—one of the last surviving officers of the battle of Trafalgar—was born at Prospect House, Broadstairs, Kent, England, in 1838. He served as midshipman on board the *Calcutta* in the Australian trade: entered the Allan service during the progress of the Crimean war, and was present at several of the engagements between the Russians and the allied forces: went to Odessa with the allied fleets, and was serving on board the *Indian* when she received sealed orders to proceed to Kinburn and lay buoys for the iron-clads which bombarded and destroyed the forts. Captain Smith's first command in the Allan service was the steamer *St. George*; subsequently he was master of the *Hibernian, Circassian, Peruvian, Sardinian* and the *Parisian*. He succeeded Captain James Wylie as Commodore of the fleet, and held that position for several years, until he resigned to accept the office of Chairman of the Board of Examiners of Masters and Mates, Commissioner for enquiring into wrecks, and one of the nautical advisers of the Government. This office he still holds with headquarters in Halifax, N.S. Capt. Smith was always very popular with the travelling community. On leaving the service he was presented with a valuable set of plate.

Alexander Aird, previous to joining the Allan

Line, had been in command of the *John Bell* and *United Kingdom* of the Anchor Line. His first command in the Allan Line was the *St. George* in 1864. Subsequently, he was captain of the *St. David*, *Nova Scotian*, *Nestorian*, *Scandinavian*, and, finally,

CAPTAIN ALEX. AIRD.

of the *Sarmatian*. Of the last-named ship he was very proud, and it was a feather in his cap that he brought out the Marquis of Lorne and Princess Louise in 1878, receiving from them a handsome recognition of his efforts to secure their comfort.

Owing to impaired health he retired from the sea some years previous to his death, which took place in 1892.

Robert Brown, of the *Polynesian*, "the rolling Polly," as she used to be called, was the *beau ideal* of

CAPTAIN RITCHIE.

a fine old English gentleman, than whom none could more gracefully discharge the honours of the table. He had many encounters with field ice off the coast of Newfoundland, but by dint of his caution, skill and patience, he invariably came out scatheless, though not unfrequently locked up in the ice for weeks at a time.

William Richardson, of the *Nova Scotian* and the *Sardinian*, who died not long ago, was an easy-going, kindly-disposed man, and a general favourite. Neil Maclean, of the third *Canadian*, was a man of fine presence and good address. Captain Joseph Ritchie, who retired from the command of the *Parisian* in 1895, though not to be called an old man, had spent forty-four years at sea. He was captain of the *Peruvian* in 1882, when the twenty-five-foot channel through Lake St. Peter was inaugurated; and again in 1888, in the *Sardinian*, he was the first to test the increased depth to twenty-seven and a half feet. Ritchie's whole career was a most successful one. On retiring from the service he was presented with a very handsomely engrossed address and a valuable service of silver plate by his Montreal friends.

Joseph E. Dutton, best known as the captain of the *Sardinian*, was a remarkable man, and frequent voyages with him led me to know him better than some of the others. "Holy Joe," as he was familiarly called, was an excellent sailor, but had to contend with a good many difficulties. At one time his ship lost her rudder in mid-ocean; at another time she lost her screw. Once she caught fire in Loch Foyle from an explosion of coal gas, and had to be scuttled. Dutton was a clever, well-read man, and a born preacher. When he had on board some eighteen clergymen going to the meeting of the Presbyterian Council at Belfast, he came into the saloon on a

Saturday evening, and coolly announced that if they had no objections he would conduct the Sunday service himself. And preach he did. He had the whole Bible at his finger-ends. I recall at least one voyage when he personally conducted three religious

CAPTAIN JOSEPH E. DUTTON.

services daily—one at 10 o'clock a.m., for the steerage passengers; one at 4 p.m., in the chart-room, and one at 7 p.m., in the forecastle, for his sailors. As to creed, he had drifted away from his early moorings, and admittedly had difficulty in finding secure anchorage. He had, so to speak, boxed the ecclesiastical compass. He had been a Methodist, a Baptist, a Plymouth Brother, but with none of them did he long remain in fellowship. Finally, he pinned his faith to the tenets of "conditional immortality," arguing with great ingenuity and earnestness that eternal life is the exclusive portion of the righteous, and annihilation that of the wicked. One of Captain Dutton's last public appearances in Montreal was on a Sabbath evening, in the Olivet Baptist church, when he baptized seven of his sailors by immersion in the presence of a crowded assemblage. He was a square-built, powerful Christian. The way he collared these men and submerged them was a caution. He gave each of them in turn such a drenching as they will remember for a long time, and all with the greatest reverence; nor did he let them go until he received from each a solemn assurance that he would be a faithful follower of Christ to his life's end. Not long after this, Captain Dutton had an attack of Bright's disease, which brought him to an early grave. He was buried in Mount Royal cemetery, where the monument, "erected by a few of his friends," bears the inscription:

"Commodore Allan Line. Lieut. R. N. Reserve. In memory of Captain Joseph E. Dutton, late of the R. M. SS. *Sardinian*. Born at Harrington, England, February 8th, 1828. Died at Montreal, July 6th, 1884, aged 56 years.

"'Now are we the sons of God, and it doth not yet appear what we shall be; but we know that when he shall appear we shall be like him.'—1 John iii. 2."

There was a time when profane swearing used to be indulged in freely by sea-captains and their subordinates. Happily the custom is going out of fashion, though now and then a representative from the old school may still be found. Captain Dutton was never addicted to swearing, though his temper was tried often enough. On arriving at Rimouski in 1879, after making the fastest voyage to the St. Lawrence then on record, the *Sardinian* had to lie at anchor for two mortal hours before he could get his mails landed. One hour it took the tender to get up steam, and another hour to get alongside the ship, owing to a strong easterly breeze, which brought up a lop of a sea. All this lost time Dutton rapidly paced the bridge to and fro with evident impatience. At length, when the tender was made fast, he came down and mingled with the crowd on deck, on the keen lookout for letters and newspapers, when one said to him, jokingly, "Why did you not swear at the captain of that tender?" "Oh," said he, with a pleasant smile, "he is only a farmer." The provocation had been great, but the controlling principle was greater and highly creditable to Dutton.

Apropos to the subject of swearing was the story told by a fellow-passenger—a deacon in the late Prof. Swing's congregation in Chicago. Dr. Swing had withdrawn from the Presbyterian Church, but continued to preach in a public hall or theatre, drawing immense crowds to hear him. Swing was a sensational preacher, who could extort tears or smiles from his hearers at will, and not unfrequently his random shots hit the mark. On one occasion, the deacon informed us, he overheard the remark made by one of Chicago's fastest young men to a comrade as they were leaving the place of worship after listening to a scathing discourse on the besetting sins of young men, swearing included: "Say, Jim, I'll be d——d if that is not the kind of preaching that suits me." This is a hard story, scarcely credible, but it was told in sober earnest and in a tone that indicated that in the speaker's judgment an arrow had pierced the young man's heart, and that the shocking expression just quoted was, after all, neither more nor less than his peculiar way of emphasizing the fact that he was *stricken*.

THE DOMINION LINE.

This line began in 1870 when a number of merchants, engaged in the New Orleans and Liverpool trade, formed what they styled the "Mississippi and Dominion Steamship Company, Limited," under the management of Messrs. Flinn, Main and Montgomery, of Liverpool, the agents in Montreal being Messrs. D. Torrance & Co., of which Mr. John Torrance has been for a number of years the senior partner. Their boats were to run to New Orleans in the winter and to Montreal in summer. Their first ships were the *St. Louis, Vicksburg* and *Memphis.* In 1871 they added the *Mississippi* and *Texas* of 2,822 tons. The Orleans route was soon abandoned and the Dominion Line, then so called, confined its trade to Canada, having Portland for its terminal winter port. Gradually increasing the size and speed of their steamers they entered into a lively competition for a share of the passenger traffic, and soon became formidable rivals of the Allan Line, and for a number of years shared with them in the Government allowance for carrying the Royal mails.

In 1874 they had built for them at Dumbarton the *Dominion* and *Ontario*, each 3,000 tons; in 1879 the *Montreal*, *Toronto* and *Ottawa*, of still larger dimensions, were added. They next bought the *City of Dublin* and *City of Brooklyn* from the Inman Line, and renamed them the *Quebec* and *Brooklyn*. In 1882 and 1883 they built the *Sarnia* and the *Oregon*, fine boats of about 3,700 tons each, with increased power and mid-ship saloons. In 1884 Messrs. Connal & Co., Glasgow, built for them the *Vancouver*, a very fine ship of 5,149 tons, having a speed of fourteen knots and excellent accommodation for passengers. Although she has had several minor accidents she has been, on the whole, a successful and popular ship. The most serious misfortune that befell her was in November, 1890, on her voyage to Quebec, when she encountered a furious hurricane in mid-ocean. Captain Lindall, who had been constantly on the bridge for a long time, went to his chart-room to snatch a few minutes rest, leaving the first officer on the bridge. All of a sudden the ship was thrown on her beam ends by a tremendous wave which completely wrecked the bridge and swept the chart-room, with the captain in it, into the sea. The quarter-master at the wheel was also washed overboard, and both he and

Captain Lindall were drowned. The first officer, Mr. Walsh, who had a miraculous escape, took charge of the battered ship and brought her to Quebec, where deep regret was expressed for the sad death of

CAPTAIN LINDALL.

Lindall, who was a general favourite and as good a sailor as ever stood on the bridge.

The *Labrador*, 4,737 tons, launched from the famous shipyard of Harland & Wolff, Belfast, in 1891, has also been a successful and popular ship. She combines in her construction a number of the latest improvements, and has attained a high rate of speed, with large cargo capacity and a moderate consumption of fuel. Until the arrival of the *Canada*, in October, 1896, the *Labrador* held the record for the fastest voyage from Moville to Rimouski—6 days, 8 hours. In August, 1895, she made the voyage from land to land in 4 days, 16 hours. In May, 1894, she averaged 365 knots a day, equal to fifteen knots an hour, her best day's run being 375 knots, which was regarded as great work considering the small amount of fuel consumed. In December of that year she made the run from Moville to Halifax in 6 days, 12 hours.

Up to this point, however, the business ability and enterprise of the Dominion Company had not been rewarded with financial success. For years they had to contend with the general depression of trade, the keen competition of other lines, and ruinous rates of freight. In the autumn of 1894 the managers resigned, and the entire fleet of vessels was sold to Messrs. Richards, Mills & Co., of Liverpool, at a great sacrifice. The Montreal agency remains as heretofore with Messrs. D. Torrance & Co., and under the new management the line seems to have entered upon a career of prosperity.

The casualties on the St. Lawrence route to steamers of this line have been numerous, but with a comparatively small loss of life. The foundering of the *Vicksburg*, from collision with ice, in 1875, was the most disastrous, involving the loss of forty-seven lives of passengers and crew—including the captain—and a large number of cattle. The *Ottawa* went ashore about fifty miles below Quebec in 1889 and became a total wreck. The *Idaho* was wrecked on Anticosti in 1890; the *Montreal,* on the island of Belle Isle in 1889. The *Texas* went ashore on Cape Race in a fog and became a total wreck. In September, 1895, the *Mariposa,* a beautiful twin-screw chartered steamer of 5,000 tons, was stranded at Point Amour in the Straits of Belle Isle and became a total wreck, but the passengers and crew were all saved.

It very soon became apparent that the new management of the Dominion Line was bent on a new departure. They lost no time in discarding the smaller boats and replacing them with large and powerful freight steamers having also limited accommodation for passengers. Of this type were the *Angloman** and the *Scotsman.* The latter is a fine twin-screw ship of colossal strength, 6,040 tons register, with a carrying capacity of from 9,000 to 10,000 tons of cargo, and an average speed at sea of twelve to thirteen knots. In September, 1895, in addition to a large general cargo, the *Scotsman* left Montreal with the largest shipment of live stock that ever left this port, consisting of 1,050 head of cattle, 2,000 sheep, and 47 horses, all of which were landed safely in Liverpool. But the latest addition to the fleet is in advance of the *Scotsman.* The *Canada,*

---

* The *Angloman* was wrecked on the Skerries, in the Irish Sea, in February, 1897. The crew were rescued, but the ship, with her valuable cargo and a large number of cattle, became a total loss, though fully covered by insurance.

DOMINION LINE SS. "CANADA."

THE ST. LAWRENCE ROUTE. 227

which sailed on her first voyage from Liverpool on October 1st, 1896, is a type of ocean steamer new to the St. Lawrence, and is designed to meet present requirements by combining in one vessel the essential

CAPTAIN MACAULAY, OF SS. "CANADA."

features of a first-class passenger ship with so large a freight-carrying capacity as to make her practically independent of subsidies. The *Canada* is a twin-screw steamer 515 feet long, 58 feet beam, and

35 feet 6 inches moulded depth. Her gross tonnage is about 9,000 tons. Her triple expansion engines are calculated to develop 7,000 horse-power with a steam boiler pressure of 175 pounds. Her staterooms are perhaps the finest feature of the ship—equal to any on the ocean ferry. Her maiden voyage was a stormy one, but it easily surpassed all previous records from Liverpool to Quebec. On her second trip she left Liverpool at 5 p m. on October 29th, and reached Rimouski on November 4th, at 11.40 p.m., thus making the voyage in 6 days, 11 hours and 40 minutes, and to Quebec in 6 days, 23 hours, 30 minutes. Her average speed on this voyage was about 16 knots an hour, and her best day's run, 416 knots, equal to $17\frac{1}{3}$ knots an hour.

At a luncheon given on board the *Canada* to leading members of the Dominion Government, Mr. Torrance said that the Dominion Line had been sold out to a company composed of men of tremendous energy and enterprise, with any amount of money at their backs, and, after looking at the matter in all its bearings, they decided that the time had come for a forward movement. They determined to build the largest steamer they could for the St. Lawrence trade. The *Canada* was contracted for by Messrs. Harland and Wolff, Belfast, as a sixteen-knot ship, and on her trial trip made seventeen and a half knots. He believed that she would average sixteen knots at sea, that she would reach Rimouski in six and a half days from Liverpool, and deliver her mails at the Montreal post-office within seven days. If that expectation comes to be realized, as it is most likely to be, the arguments in favour of a fast mail service between Canada and Britain will be materially strengthened. Mr. Torrance added that the *Canada* was built to carry 7,000 tons of cargo, that if she had a speed of seventeen knots she would only carry 4,000 tons of cargo; if eighteen knots, she would carry but 3,000 tons, and that with a speed of twenty knots it would not be safe to calculate on her capacity for more than 1,000 tons of freight; "in short, that the twenty-knot ship must be, virtually, a passenger ship, and well subsidized." The Canadian Government has not been slow to back up private enterprise of this nature in the past, and will doubtless continue to do so in the future. For reasons not made public the *Canada* was withdrawn from the St. Lawrence service and placed on the route from Boston and Liverpool, where she has been so successful that another vessel of the same class is being built for that route. In the meantime other large vessels have been put on the St. Lawrence route, the latest addition to the fleet being the *New England*, having a tonnage of nearly 11,600 tons, fine accommodation for a large number of passengers, and room for an enormous cargo.

The Beaver Line.

This is an out-and-out Canadian enterprise, dating from 1867, under the name of the "Canada Shipping Company, Limited," when several Montreal capitalists, among whom were the late William Murray and

ROYAL MAIL SS. "LAKE ONTARIO," BEAVER LINE.

Alexander Buntin, Messrs. Alexander Urquhart, John and Hugh Maclennan and others, combined to originate a line of iron fast-sailing ships to trade between Montreal and Liverpool. Having adopted for its distinguishing flag the emblem of the Canadian beaver, the company soon came to be popularly known as the Beaver Line, a line which, though not remunerative to its originators and stockholders, is worthy of honourable mention as having contributed in many ways to the interests of Canadian trade and commerce. The company commenced with a very fine fleet of five Clyde-built iron ships of from 900 to 1,274 tons each. These were the *Lake Ontario*, the *Lake Erie*, the *Lake Michigan*, the *Lake Huron* and the *Lake Superior*. The ships were in themselves all that could be desired. They were beautiful to look at, and made swift voyages, but there was a necessary element of success wanting. They did not pay. In fact, they began their short-lived career at the time when the days of sailing ships were rapidly drawing to a close. The important question of steam *versus* sails had been settled. The Canada Shipping Company must therefore retire from the business altogether or avail themselves of the advantages of steam power. They decided upon making the experiment, and gave orders for the building of steam vessels to supersede the sailing ships. In the meantime the *Lake Michigan* was lost at sea with all on board, adding another to those mysterious disappearances, of which there have been so many instances— gallant ships and noble sailors setting out on their voyage buoyant with hope, reporting themselves at the last signal station as "all well," but never to be heard of any more.

The *Lake Huron* was wrecked on Anticosti. The year 1875 saw the first steamers of the Beaver Line afloat. They were the *Lake Champlain, Lake Megantic* and *Lake Nepigon*, snug little ships of about 2,200 tons each, such as would pass nowadays for cruising steam yachts, but much too small for cargo ships on the Atlantic, to say nothing of the passenger business. The *Lake Manitoba* and *Lake Winnipeg*, of larger size and higher speed, were added in 1879, followed by the *Lake Huron* and the *Lake Superior*. The last-named is a fine ship of 4,562 tons, and credited with thirteen knots an hour. It was not long before three of the steamers came to grief. The *Lake Megantic* was wrecked on Anticosti in July, 1878; the *Lake Manitoba*, on St. Pierre Island, in the Gulf of St. Lawrence, in June, 1885; the *Lake Champlain*, stranded on the north coast of Ireland in June, 1886. To keep up the weekly line, the *Lake Ontario*, built at Sunderland in 1887, was purchased at a cost of nearly $300,000. She is a vessel of about 4,500 tons, with midship saloon, triple expansion engines, and a maximum speed of thirteen knots. She is an excellent sea boat, with good accommodation for one hundred cabin passengers. The ships of this line all carry live cattle, sheep and horses, for which they are well adapted. The Beaver Line led the way towards the reduction of transatlantic cabin passage rates on the St.

Lawrence route. It also introduced the custom of embarking and landing passengers at Montreal instead of Quebec as formerly. Unfortunately the line had not been a success financially. In the winter of 1895 the boats were all tied up, the com-

CAPTAIN HOWARD CAMPBELL.

pany went into liquidation, and the entire fleet was sold at a nominal price to the bondholders. During the following winter, however, the ships of this line maintained a weekly service from Liverpool to St. John, N.B., receiving from the Canadian Government a subsidy of $25,000, and in 1897 the Beaver Line was awarded the contract for carrying the Canadian mails, to be landed at Halifax in the winter months. The annual subsidy for this service is understood to be $146,000. This arrangement, however, is necessarily of a temporary nature, pending the development of the long-expected "fast service." In the meantime the Beaver Line has added to its fleet the fine SS. *Gallia*, of the Cunard Line, and the *Tongariro*, of 4,163 tons, formerly belonging to the New Zealand Shipping Company. The service has thus far been satisfactory.

Captain Howard Campbell, of the SS. *Lake Ontario*, died very suddenly on Sunday morning, April 3rd, 1898. The second day out from Halifax towards Liverpool, he went on the bridge, sextant in hand, intending to take an observation. While in the act of doing so he fell into the arms of a quartermaster and died instantly. Captain Campbell had been long connected with the Beaver Line. He was widely known as a skilful mariner and a genial and accomplished man. He was born at St. Andrews, N.B., and was fifty-four years of age.

There are a number of other lines of steamships plying regularly from Montreal in summer and from different Atlantic ports in winter. They are chiefly cargo and cattle ships, with limited accommodation for passengers. Among these are the Donaldson Line, with five ships of from 2,000 to 4,272 tons, giving a weekly service to Glasgow and Bristol; the

Thomson Line, with seven ships to London, Newcastle and Antwerp. The Johnston Line has regular sailings to Liverpool. The Ulster Steamship Company, or "Head Line," has five ships running to Belfast and Dublin fortnightly. The Elder, Dempster Line has a fleet of sixteen large freight steamers, ranging from 4,500 to 12,000 tons each. Some of them are fitted with cold storage, and all of them have the modern improvements for carrying live stock and grain; they maintain a regular weekly service to London and to Bristol.* The Hansa St. Lawrence Line plies to Hamburg and Antwerp; the Furness Line to Antwerp and Dunkirk, and also to Manchester.† The Quebec Steamship Company has regular communication with Pictou, N.S., by the fine upper saloon steamship *Campana*, of 1,700 tons. The Black Diamond Line has five ships of from 1,500 to 2,500 tons each, plying regularly in the coal trade from Montreal to Sydney, Cape Breton, Charlottetown, P.E.I., and Newfoundland.

*The SS. *Memphis*, of the African Steamship Company, but employed by the Elder, Dempster Line, went ashore on the west coast of Ireland in a fog in November, 1896, and became a total wreck. Ten of the crew were drowned and 350 head of cattle.

†The Manchester ship canal is 35 miles long, 120 feet bottom width, and 26 feet in depth. The docks at Manchester cover 104 acres and have five miles of quays. It was estimated to cost £10,000,000 sterling, but cost over £15,000,000 before it was completed. Arrangements are in progress by a Manchester syndicate for the establishment of a weekly line of steamships of 8,500 tons capacity, to be provided with cold storage and the most approved equipments for carrying live stock. The best modern appliances for loading and discharging cargo, grain elevators being included, are among the attractions which enterprising Manchester presents to the shipping trade of Canada.

The export trade in live stock, which commenced here in 1874 with only 455 head of cattle, has now assumed large proportions. In 1897 there were shipped from Montreal 119,188 head of cattle, 12,179 horses and 66,319 sheep, valued in all at about $8,700,750. The cattle were valued at $60 a head, the horses at $100, and the sheep at $5.00 each. The ocean freight on cattle was $10 per head, and on sheep $1.00 each.*

Canadian Fast Atlantic Service.

Ever since the completion of the Canadian Pacific Railway in 1885, the idea of instituting a fast service between Great Britain and the St. Lawrence has been regarded with yearly increasing favour. Now it is regarded as a necessary link in the chain that binds the colony to the Mother Land, and indispensable if this route is to become Britain's highway to the East.

As early as 1887 the Canadian Government advertised for tenders for a line of Atlantic mail steamers to have an average speed of 20 knots an hour, coupled with the condition that they should touch at some French port. The Allans, who at that time deemed a 20-knot service unsuited to the St. Lawrence route, offered to supply a weekly service with a guaranteed average speed of 17 knots, for an annual subsidy of $500,000 on a ten years' contract. That offer was declined. About the same time the English firm of

* "Montreal Board of Trade Report, 1897," pp. 52, 88.

Anderson, Anderson & Co. offered to provide a line of vessels "capable of running 20 knots" for the same subsidy. This dubious offer was accepted provisionally by the Canadian Government, but it was eventually fallen from. Two years later another abortive attempt was made, when the Government of the day voted $750,000 as an annual subsidy for a 20-knot service; but nothing resulted. In 1894 Mr. James Huddart, of Sydney, N.S.W. (the contractor for the Vancouver-Australian Line of steamers), entered into an agreement with the Dominion Government for a weekly 20-knot service for said amount of $750,000 per annum. For reasons that need not be explained, this proposal also fell through. In 1896 the Allans were said to have tendered for a 20-knot service on the basis of a subsidy of $1,125,000, but the offer was declined owing to some informalities.

In view of so many failures it is scarcely safe to affirm that the fast service is now assured. In May, 1897, however, it was officially announced by the Canadian Government that a contract had been entered into, with the approval of the British Government, whereby Messrs. Peterson, Tate & Co., of Newcastle-upon-Tyne, agreed to furnish a weekly service with a guaranteed speed of at least 500 knots a day. The contractors are to provide four steamers of not less than 520 feet in length, with a draft of water not exceeding 25 feet 6 inches. The ships are to be not less than 10,000 tons register, fitted to carry from 1,500 to 2,000 tons of cargo, with suitable cold storage accommodation for at least 500 tons. They are to be equal in all respects to the best Atlantic steamships afloat, such as the *Campania* and *Lucania*, with accommodation for not less than 300 first-class, 200 second-class and 800 steerage passengers. The annual subsidy is to be $750,000, whereof the Canadian Government is to pay $500,000 and the British Government $250,000. The steamers are not to call at any foreign port, and the company is forbidden to accept a subsidy from any foreign country. The mails are to be carried free. The termini of the line will be Liverpool and Quebec during summer, the ships proceeding to Montreal if and when the navigation permits. In winter the Canadian terminus will be Halifax or St. John, N.B., at the option of the contractors, who are to provide a 22-knot tender of the torpedo type to meet each steamer on her approach to the Canadian coast when required, and pilot her to her destination. The contractors must deposit £10,000 in cash, and a guarantee of £10,000 additional, with the Minister of Finance of Canada as security that the contract will be faithfully carried into effect.

Twelve months having passed since the signing of the contract, without any substantial progress having been made towards its fulfilment, a new agreement was entered into in April last whereby the Government granted Messrs. Peterson and Tate an extension of time, and introduced several important changes into the contract. Under the new arrangement the contractors were required to have a steamship company incorporated by May 30th, 1898, with a substantial capital of $6,250,000, to have contracts signed with ship-builders at that date for four steamships, and to have two of them actually under construction. The 1st of May, 1900, was named as the time when the four steamers are to be ready to go on the route and commence a regular weekly service. The preliminary conditions attached to the contract appear to have been complied with, and a company has been incorporated under the name of the "Canadian Royal Mail Steamship Company, Limited;" but grave fears are entertained that the necessary funds may not be forthcoming, and that the long-expected fast service may be indefinitely delayed.

Sir Sandford Fleming, who has made a study of this subject, and published his opinions respecting it in a series of pamphlets, is not sanguine as to the success of the undertaking. "The conditions imposed by nature," he says, "are unfavourable for rapid transit by the St. Lawrence route, and any attempts to establish on this route a line of fast transatlantic steamships to rival those running to and from New York would result in disappointment." In the event of such a service being instituted, Sir Sandford assumes that it would be almost exclusively for the use of passengers, and suggests that the route should be from Loch Ryan, on the Wigtonshire coast of Scotland, to North Sydney, in Cape Breton. The distance between these points being only 2,160 knots, the voyage might be made in $4\frac{1}{2}$ days, while 30 hours more would land mails and passengers in Montreal by railway. In this way the average time from

London to Montreal would be reduced to 6 days and 6 hours—36 hours less than the time usually occupied between Montreal and London *via* New York and Queenstown.

"In connection with the ocean service there might also be a line of fast light-draught steamers to run to and from Montreal to Sydney and the Gulf ports. In this way the people of the Maritime Provinces, including Newfoundland, would share in the benefits to be derived from the fast ocean service equally with those of Quebec and Ontario." Sir Sandford's idea is to have the fastest ocean ship on the shortest ocean passage, and by all means to avoid the Straits of Belle Isle, "the saving of a few hours being insufficient to counterpoise the tremendous risks to which fast passenger steamships, in navigating the Belle Isle route, would so seriously and frequently be exposed." It is claimed that if this plan were adopted three ocean steamers would suffice instead of four. Reference to the accompanying sketch-map, showing the relative positions of Sydney, Newfoundland, and the Straits of Belle Isle, with the existing lines of railway, will help to make Sir Sandford's proposal clear.

Among other proposals, an English syndicate recently offered to furnish a 24-knot service between Milford-Haven, on the coast of Wales, and a port in Nova Scotia, representing to the British Government that they would be able to carry troops across the Atlantic in four days, and land them in Victoria in six days more. But the 24-knot steamship has not yet been launched.

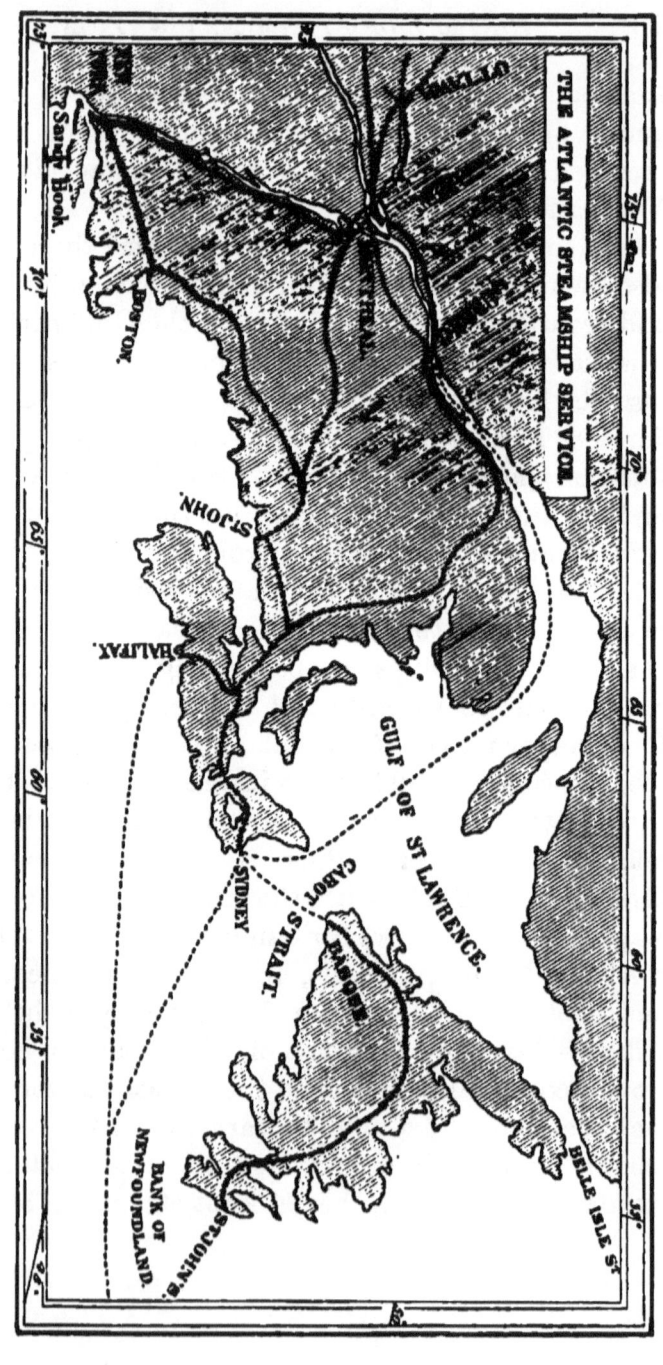

MAP OF THE GULF OF ST. LAWRENCE AND NORTH ATLANTIC PORTS.
(Kindly furnished by Sir Sandford Fleming.)

Sir Sandford Fleming, K.C.M.G., LL.D., C.E., is one of Canada's most eminent civil engineers. He was born at Kirkcaldy, Fifeshire, Scotland, January 7th, 1827, came to Canada at the age of eighteen, and has ever since been identified with the progress and development of the country. He was on the engineering staff of the Northern Railway from 1852 to 1863, and for the latter half of that time was chief engineer of the work. He was chief engineer of the Intercolonial Railway, and carried it through to a successful completion in 1876. In 1871 he was appointed engineer-in-chief of the Canadian Pacific Railway; he retired from that position in 1880 and was subsequently elected a director of the company. He received the freedom of the Royal Burgh of Kirkcaldy and the degree of LL.D. from the University of St. Andrews in 1884: was appointed to represent Canada at the International Prime Meridian Conference in Washington in 1884: at the Colonial Conference, London, in 1887, at the Colonial Conference in Ottawa, in 1894, and at the Imperial Cable Conference in London, in 1896. Sir Sandford has been Chancellor of Queen's University at Kingston since 1880. He is the author of numerous scientific and other publications, is an active member of the Royal Colonial Institute of London, and on the occasion of Her Majesty's Diamond Jubilee was accorded the honour of knighthood.

The conflicting rumours, which for many months have been in circulation as to the inability of Messrs.

Peterson, Tate & Company to fulfil the terms of their agreement, have finally been set at rest by the cancelling of the contract, and the Canadian Government calling for tenders for a weekly steamship service for carrying Her Majesty's mails for a period of two years from the 1st of May, 1899, from Montreal and Quebec to Liverpool, during the summer months, and from St. John, N.B., and Halifax in winter. The time occupied in making the voyage from Rimouski to Moville and *vice versa*, is not to exceed an average of seven days. This is clearly a temporary arrangement and not an implied abandonment of a faster service than already exists. The opinion, however, in business circles seems to be gaining ground that something much less costly than a twenty-knot service might for some years to come meet the requirements of the country.

CHAPTER VIII.

STEAM ON THE GREAT LAKES.

The History of Steam Navigation on the Great Lakes—The Construction of the St. Lawrence, the Welland, and the Rideau Canals—The Port of Montreal.

THE waterways of Canada available for steam navigation are on a magnificent scale. The main system extends from the mouth of the St. Lawrence at Belle Isle to Fort William and the head of Lake Superior—a distance of nearly 2,384 miles, and rendered navigable without interruption by a series of ship canals. Proceeding four hundred miles farther west, another long stretch of inland navigation begins with Lake Winnipeg, 240 miles long, into which, at its northern extremity, flows the mighty Saskatchewan, navigable for steamers one thousand miles! Not to mention smaller streams, the rivers St. John and Miramichi, in the Province of New Brunswick, afford 300 miles of navigable water and float a large amount of shipping. Ships of the largest size can ascend the Saguenay seventy-five miles. The Ottawa in its several reaches is navigable by steam for three or four hundred miles. Steamers ply on the Assiniboine, 250 or 300 miles westward from Winnipeg. The Mackenzie River is navigable for a thousand miles. The Fraser, the Thompson, and the Columbia rivers in British Columbia contribute largely to the steam tonnage of the Dominion. The Great Lakes,* commonly so called, are in reality great inland fresh water seas, often swept by gales of wind and combing billows, and sometimes, alas, strewed with wrecks. They have their breakwaters, lighthouses and steam fog-signals as fully equipped as similar oceanic structures and appliances. The Lake of the Woods and Lake Manitoba are each 100 miles long.

As early as the year 1641 a few Jesuit missionaries and fur-traders had reached the rock-bound shores of Lake Superior in their canoes, but it is not until some years later that history makes us acquainted with the first sailing vessels that appeared on the lakes. One of the earliest of these was a schooner of ten tons, built near where Kingston now is by the enter-

* DIMENSIONS OF THE GREAT LAKES.

LAKES.	Length. (Miles.)	Greatest Width. (Miles.)	Depth. (Feet.)	Above Sea. (Feet.)	Area. (Sq. Miles.)
Ontario	180	65	500	247	7,300
Erie	240	80	210	573	10,000
Huron	280	190	802	581	24,000
† Michigan	335	88	868	581	25,600
Superior	420	160	1,008	601	32,000

† Lake Michigan lies wholly within the United States.

prising French adventurer, La Salle, who had been appointed Governor of Fort Frontenac, and had a roving commission to explore the western wilds of North America. Accompanied by the famous Recollet Father, Hennepin, and some thirty others, La Salle set sail on the 10th of November, 1678, for the head of Lake Ontario. Finding his further passage barred by the Falls of Niagara, he wintered in that neighborhood and had another vessel built at Cayuga Creek, a few miles above the Falls. This vessel, which he named the *Griffin*, of about sixty tons burthen, was launched in May, 1679, and was probably the first to navigate the upper lakes. On the 7th of August the *Griffin*, equipped with seven guns and a diversity of small arms and freighted with a load of goods, sailed away for Detroit and parts unknown. The Detroit River was reached in a few days, and Green Bay—at the head of Lake Huron—some time in September, when she was loaded with furs and despatched on her return voyage to Niagara, which, however, she never reached, the vessel and cargo having been totally lost on the way. For many years after the loss of the *Griffin* the navigation of the lakes seems to have been chiefly confined to bateaux, and up to 1756 the construction and use of sailing vessels was largely, if not entirely, confined to Lake Ontario. The first American vessel built on Lake Erie was the schooner *Washington*, built near Erie, Pa., in 1797. After plying on Lake Erie one season, she was sold to a Canadian and carried on wheels around the Falls to Lake Ontario, where she sailed from Queenston for Kingston in 1798 as a British vessel, under the name of *Lady Washington*. In 1816 the whole sailing tonnage on Lake Erie was only 2,067 tons. In 1818 the fleet on Lake Ontario numbered about sixty vessels.

It is not necessary to enlarge on the growth and decadence of sailing vessels on the Great Lakes. Suffice it to say that the sailing vessel had reached its palmiest days between the years 1845 and 1862. In the latter year the gross tonnage of the lakes had risen to 383,309 tons, valued at $11,865,550, and was divided as follows: 320 steamers, aggregating 125,620 tons; and 1,152 sailing vessels, aggregating 257,689 tons. Side-wheel steamers numbered 117, and propellers, 203. In 1896 the entire number of sailing vessels on the Northern Lakes (including Lake Champlain) was 1,044, and of steam vessels, 1,792. Many in both of these classes were small vessels, including yachts and barges: the number actually engaged in the transportation business would be about 774 sailing vessels and 1,031 steamers over fifty tons burthen — a large proportion of the steamers being from 1,500 to 2,500 tons burthen.*

Coming back now to the beginning of steam navigation on the Great Lakes, we find that the first Canadian steamer to navigate any of these waters was the *Frontenac*, built at Finkle's Point, eighteen miles above Kingston, by Teabout & Chapman, of

---

* These figures refer exclusively to vessels belonging to the merchant marine of the United States on the Great Lakes and are taken from official reports.

Sackett's Harbour, for a company of shareholders belonging to Kingston, Niagara, Queenston, York and Prescott. The *Frontenac* was launched on September 7th, 1816. Her length over all was 170 feet, and her registered tonnage, 700 tons. She cost nearly £20,000 currency. The engines were made by Watt & Boulton, of Birmingham, England, and cost about £7,000. The *Frontenac* was said to be the best piece of naval architecture then in America, and her departure on her first voyage was considered a great event—" she moved off from her berth with majestic grandeur, the admiration of a great number of spectators." Her maiden trip for the head of the lake was commenced on June 5th, 1817. Her regular route was from Prescott to York (Toronto) and back, once a week. She was commanded as long as she was afloat by Captain James Mackenzie, a gallant sailor who had previously served in the Royal navy. The *Frontenac* eventually became the property of the Messrs. Hamilton, of Queenston. She was maliciously set on fire by some miscreants while lying at her wharf at Niagara in 1827, and was totally destroyed.

About the same time the Americans had built a steamboat at Sackett's Harbour, N.Y., named the *Ontario*, a vessel 110 feet long, 24 feet wide, and $8\frac{1}{2}$ feet in depth, measuring 240 tons. The *Ontario* made her first trip in April, 1817, thus establishing her claim of precedence in sailing on the lakes. She was built under a grant from the heirs of Robert Fulton. On her first trip she encountered considerable sea, which lifted the paddle-wheels, throwing the shaft from its bearings and destroying the paddle-boxes. This defect in her construction having been remedied, she was afterwards successful, it is said, but her career is not recorded.* The Americans built another steamer at Sackett's Harbour in 1818, the *Sophia*, of 70 tons, to run as a packet between that port and Kingston. In that year also the Canadians built their second lake steamer, the *Queen Charlotte*. She was built at the same place as the *Frontenac*, and largely from material which had not been used in the con-

"QUEEN CHARLOTTE."
Second steamer on Lake Ontario, 1818.

struction of that vessel. She was launched on the 22nd of April, 1818, and was soon ready to take her place as the pioneer steamer on the Bay of Quinte.†

---

* Mr. C. H. Keep, in his report on the "Internal Commerce of the United States for 1891," has given a graphic History of Navigation on the Great Lakes, and is our chief authority for these notes on the early American lake steamers.

† Robertson's "Landmarks of Toronto," p. 847.

The *Queen Charlotte* was a much smaller boat than the *Frontenac*. Her machinery was made by the brothers Ward, of Montreal, and she seems to have plied very successfully for twenty years from Prescott to the "Carrying Place" at the head of the Bay of Quinte, where passengers took stage to Cobourg and thence proceeded to York by steamer. She was com-

"WALK-IN-THE-WATER."
First steamer on Lake Erie, 1818.

manded at first by old Captain Richardson, then for a short time by young Captain Mosier, and afterwards, to the end of her career, by Captain Gildersleeve, of Kingston. She was finally broken up in Cataraqui Bay; but in the meantime upwards of thirty steamers were plying on Lake Ontario and the Upper St. Lawrence, to some of which particular reference will be made later on.

The first steamer on Lake Erie was the *Walk-in-the-Water*, built at Black Rock, near Buffalo, by one Noah Brown, and launched May 28th, 1818. She was schooner-rigged, 135 feet in length, 32 feet beam and 13 feet 3 inches deep: her tonnage was $383\frac{6\,0}{9\,5}$ tons. Her machinery was brought from Albany, a distance

THE "VANDALIA."

From *Scribner's Magazine* for March, 1890.

of three hundred miles, in wagons drawn by five to eight horses each. She left Black Rock on her first voyage August 25th, and reached Detroit, 290 miles, in 44 hours 10 minutes. "While she could navigate down stream, her power was not sufficient to make headway against the strong current of the Niagara River. Resort was therefore made to what was known in the early days as a "horned breeze." The *Walk-in the-Water* was regularly towed up the Niagara River by a number of yokes of oxen, but once above the swift current she went very well. She made regular trips between Black Rock and Detroit, occasionally going as far as Mackinac and Green Bay on Lake Huron, until November, 1821, when she was driven ashore near Buffalo in a gale of wind and became a total wreck. Her engines, however, were recovered and put in a new boat named the *Superior*, in 1822. Soon after this the first high-pressure steamer on the lakes was built at Buffalo. She was named the *Pioneer*. In 1841 the first lake propeller was launched at Oswego. This was the *Vandalia*, of 160 tons, said to be the first freight boat in America to make use of Ericsson's screw propeller. She made her first trip in November, 1841, and proved entirely successful. In the spring of 1842 she passed through the Welland Canal, and was visited by large numbers of people in Buffalo, who were curious to see this new departure in steam navigation, and the result was that two new propellers were built in that year at Buffalo, the *Sampson* and the *Hercules*.

Soon after the introduction of steamboats, and because of them, when as yet railroads were not in this part of the world, Lake Erie became the great highway of travel to the western States, and it was not long until magnificent upper cabin steamers, carrying from 1,000 to 1,500 passengers, were plying between

Buffalo and Chicago. The writer well remembers making the voyage in one of these steamers late in the autumn of 1844, and that, owing to the tempestuous state of the weather, we had to tie up most every night, so that the voyage lasted nearly a whole week. The crowd of passengers was great, but it was

THE "PRINCETON."

First propeller on the lakes that had an upper cabin—one of a fleet of fourteen passenger steamers plying between Buffalo and Chicago in 1845— had twin screws, and a speed of eleven miles an hour.

a good-natured crowd, bent on having a "good time." Dancing was kept up in the main saloon every evening till midnight, after which many of us were glad to get a shake-down on the cabin floor.

The year 1836 marks an important era in the navigation of the Great Lakes, for in that year the first cargo of grain from Lake Michigan arrived at Buffalo, brought by the brig *John Kenzie* from Grand River. It consisted of three thousand bushels of wheat. Previous to that date the commerce of the lakes had been all westward, and, curiously enough, the cargoes carried west consisted for the most part of flour, grain and other supplies for the new western settlements. In 1840 a regular movement of grain from west to east had been established.

In the early years of the grain trade the loading and unloading of vessels was a very slow and irksome business. As much as two or three days might be required to unload a cargo of 5,000 bushels. In the winter of 1842-43 the first grain elevator was built at Buffalo, and a new system of handling grain introduced which was to prove of incalculable benefit to the trade. The schooner *Philadelphia*, of 123 tons, was the first to be unloaded by the elevator.

The Canadian steam traffic on Lake Erie commenced with the steamers *Chippewa* and *Emerald*, plying between Chippewa and Buffalo; the *Kent*, which foundered in 1845; the *Ploughboy*, owned by a company in Chatham, and the *Clinton*, owned by Robert Hamilton, of Queenston. A much larger Canadian steam traffic developed on Lake Huron. One of the earliest passenger steamers on the Georgian Bay was the *Gore*, of 200 tons, built at Niagara in 1838, and called after the Lieutenant-Governor of that name. That boat, which had plied for some years between Niagara and Toronto, was placed on the route between Sturgeon Bay and Sault Ste. Marie. On Lake Huron proper, the *Bruce Mines* was probably the earliest

Canadian steamer. She was employed in carrying copper ore from the Bruce mines to Montreal, and was wrecked in 1854. Shortly after, on the completion of the Northern Railway, in 1854, the company, with a view to developing their interests, entered into a contract with an American line of steamers to run from Collingwood to Lake Michigan ports tri-weekly and

THE "EMPIRE."

Built at Cleveland in 1844; a notable steamer in her day, being the largest, the fastest, and the most handsomely fitted-up vessel on the Upper Lakes at that time; ran many years between Buffalo and Chicago.

once a week to Green Bay. In 1862 six large propellers were put on the route. Later, a line of first-class passenger steamers began to ply twice a week from Collingwood and Owen Sound to Duluth at the head of Lake Superior. Among the steamers of that line, which became very popular, were the *Chicora, Francis Smith, Cumberland,* and *Algoma.* These in turn were superseded by the magnificent steamers of the Canadian Pacific and other lines elsewhere referred to.

The commerce of Lake Superior developed long after that of the lower lakes had been established. In the earliest records of the navigation of this lake, a brigantine named the *Recovery*, of about 150 tons, owned by the North-West Fur Company, is mentioned as being one of the first to sail on Lake Superior, about the year 1800. It is said that during the war of 1812, fearing that she might be seized by the Americans, her spars were taken out and her hull was covered up by branches and brushwood in a sequestered bay till peace was proclaimed. She was then taken from her hiding-place and resumed her beat on the lake until about 1830, when she was run over the Sault Ste. Marie rapids and placed in the lumber trade on Lake Erie, under the command of Captain John Fallows, of Fort Erie, Canada West. Another vessel, the *Mink*, is mentioned as having been brought down the rapids at an earlier period. In 1835 the *John Jacob Astor*, accounted a large vessel in her time, was built on Lake Superior for the American Fur Company, and placed in command of Captain Charles C. Stanard, who sailed her until 1842, when Captain J. B. Angus became master and remained in charge of her until she was wrecked at Copper Harbour in September, 1844. Passing by a number of other sailing vessels we come now to the introduction of steam on Lake Superior, and this, according to the statement of an old resident at Fort William, is how it began.

The twin-screw propeller *Independence*, Captain A. J. Averill, of Chicago, was the first steamer seen on Lake Superior. This vessel, rigged as a fore-and-aft schooner, was about 260 tons burthen, and was hauled over the Sault Ste. Marie rapids in 1844. Her route of sailing was on the south shore of the lake. Another propeller, the *Julia Palmer*, was in like manner dragged up the Ste. Marie rapids in 1846, and was the first steamer to sail on the north shore. At intervals, prior to the opening of the ship canal, several other steamers were taken up the rapids, among which were the propellers *Manhattan*, *Monticello*, and *Peninsular*, and the side-wheel steamers *Baltimore* and *Sam Ward*.

Previous to the completion of the Welland Canal the transportation of freight over the portage from Queenston to Chippewa had come to be quite a large business, giving employment to many "teamsters," for the entire traffic between Lake Erie and Ontario at this point was by means of the old-fashioned lumber-wagon. At the Sault Ste. Marie portage, Mr. Keep informs us that "one old grey horse and cart" did the business for a time, but as the volume of trade increased two-horse wagons were employed until 1850, when a light tram-road was built by the Chippewa Portage Company, operated by horses, which with a capacity for moving three or four hundred tons of freight in twenty-four hours, answered the purpose up to the time of the opening of the canal in 1855.

The Canadian Canals.

Before the construction of canals these great inland waters were of but little value to commerce, the only means of reaching them being by the bark canoe or bateau of the voyageur. The United Empire Loyalists who came to Canada at the close of the American war were conveyed to their settlements on the St. Lawrence and Bay of Quinte in the long sharp-pointed, flat-bottomed boats of the period, called "bateaux," by a very slow, laborious and uncomfortable process. General Simcoe, the first Lieutenant-Governor of Upper Canada (1791-96), is said to have sailed from Kingston to Detroit in his bark canoe, rowed by twelve chasseurs of his own regiment and followed by another canoe carrying his tents and provisions. Many still living recollect how Sir George Simpson, Governor of the Hudson's Bay Company, made his annual canoe journeys from Montreal to the Red River country. Having "sung at St. Ann's their parting hymn," his flotilla of canoes ascended the Ottawa, breasted the rapids, and by river, lake and portage, after many weary days, reached Lake Huron and the Sault Ste. Marie, thence along the north shore of Lake Superior to Fort William and the Grand Portage and by Rainy Lake and Lake of the Woods to Fort Garry. "With the self-possession of an emperor he was borne through the wilderness. He is said to have made the canoe journey to the Red River *forty times*. For his distinguished management of the Hudson's Bay Company's affairs and for his services to the trade of Canada, Governor Simpson was knighted. He died in 1860, a man who would have been of mark anywhere."*

As early as A.D. 1700 a boat canal was constructed by the Sulpicians to connect Lachine with Montreal *via* the Little St. Pierre River. The depth of water was only two and a half feet. About the year 1780 certain short cuttings with locks available for canoes and bateaux were made at a few points on the St. Lawrence where the rapids were wholly impassable. About the beginning of the century the Government of Lower Canada, appreciating the advantages of improved navigation, made liberal appropriations to that end, resulting in the completion, in 1804, of a channel three feet in depth along the shore line of the Lachine Rapids connected with short canals at the Cascades, Split Rock, and Coteau du Lac, which were provided with locks eighty-eight feet long and sixteen feet wide—small dimensions, perhaps, but at the time regarded as a vast improvement, admitting of the passage of "Durham boats," which then took the place of bateaux, with ten times their capacity. Two small locks had also been built at the Long Sault rapids, above Cornwall. But at many points the aid of oxen and horses was required, and for many years, up to the opening of the St. Lawrence canals, indeed, the chief cash revenues of the farmers along the river front were derived from the towage of barges up the swift water, in many cases to the serious neglect of their farms. In the

* Bryce's "Short History of the Canadian People," p. 333.

spirit of the religion of the early voyageurs and boatmen, crosses were erected at the head of the rapids, suggesting to those who had successfully surmounted them to rest and be thankful; hence the name, still applied to the district immediately above the Long Sault rapids, "Santa Cruz." Here, no doubt, stood for many years one of the holy crosses before which, on bended knee, thanks would often be given for a safe ascent of the rapids.

The mail service in these days between Montreal and Kingston was in keeping with the times. It was undertaken by a walking contractor, who with the mail on his back took up his line of march from Montreal, gauging his speed to accomplish the walk to Kingston and return in fourteen days.*

A good many years later it was a four days' journey from Montreal to New York by the most expeditious route then existing. Thus it was advertised in the Montreal *Gazette*, November 25th, 1827:

> DAILY STAGES. ALBANY AND MONTREAL LINE. SEASON OF 1826 AND 1827. The only full and perfect line running between Montreal and Albany leaves B. Thatcher's office, No. 87 St. Paul Street, Montreal, every day, passing through Laprairie, Burlington, Middlebury, Poultney and Salem to Albany, through an old-settled, rich and populous country, and mostly on a smooth gravelly turnpike. Through in three days, and fare very reasonable. Extras and expresses at a moment's notice. Young, Swain, Esinhart and others, proprietors.

The voyage of the Durham boat was a very tedious one, depending as it did largely on a favouring easterly breeze in traversing the lakes and quieter portions

---

* Hugh McLennan's "Lecture on Canadian Waterways, 1885."

of the river, and on the dexterity of the boatmen who wielded the "setting-poles"* in swifter water, as well as their *luck* in surmounting the rapids, where they were liable to be detained for hours, sometimes for days, contending against the swift currents, subject to the mishaps of grounding or being damaged by big boulders, or, worse still, of being caught by an eddy or an out-current and swept down the rapids, sometimes with the loss of the oxen or horses which had them in tow, and in some instances with the loss of the boat and cargo. Woe to the teamster who was not provided with a knife to cut the rope in such an emergency!

The first Lachine Canal proper, for barges, was commenced July 17th, 1821, and was completed in 1825, at a cost of $438,404. Of this amount $50,000 was contributed by the Imperial Government on condition that all military stores should be free from toll. It had 7 locks, each 100 feet long, 20 feet wide, and with 4½ feet depth of water on the sills. In 1843-49 it became a "ship canal" with 5 locks, each 200 feet

---

* The setting-pole might be twenty-five feet long, heavily shod with iron at one end and at the other fitted with a rounded knob. This pole was dropped into the water at the bow of the boat, and the boatman having put his shoulder to the other end of it, facing the stern, and pushing with all his might, walked to the farther end, cleats being fastened to the deck to give him foothold. By the time he reached the stern the barge had advanced exactly its own length, when he withdrew the pole, dragged it to the bow and repeated the process. Two or three men on each side of the boat would be similarly employed, and so the barge dragged its slow length along, much after the fashion of the horse-boat, only that the horse tugged at a stationary post while the men pushed from it.

long, 45 feet wide, and 9 feet depth of water, costing $2,149,128. The recent enlargement, commenced in 1875, cost $6,500,000. By this the locks were increased to 270 feet in length and 14 feet depth of water throughout the canal.

The Welland Canal.

The necessity of devising means to overcome the stupendous obstacle to navigation caused by the Falls of Niagara had long been apparent, but it was not until 1824 that work was commenced on the Welland Canal which was to connect Lake Ontario with Lake Erie and the west. This important work was completed in 1829, chiefly through the energy and perseverance of the Hon. William Hamilton Merritt, son of a U. E. Loyalist family, born in New York State in 1793. A man of great enterprise, he had this project on the brain for years, but like Cunard and his steamships, had difficulty in "raising the wind" —the people and the Government of Upper Canada being at that time both alike poor. He crossed the Atlantic, and, on the ground of military expediency, was said to have secured a subscription of £1,000 from the Duke of Wellington, which greatly aided him in the formation of a joint stock company who carried the work to its successful completion. The original locks were constructed of wood, 120 feet in length, 20 feet wide, with $7\frac{1}{2}$ feet of water on the sills. The entire length of the canal was twenty-six miles. This accommodated vessels carrying 5,000 bushels of wheat. Half a million of pounds were spent upon it up to the year 1841, when it was assumed by the United Canadas* and immediate steps taken for its enlargement. With locks 145 x 26 x 9, vessels loaded with 20,000 to 23,000 bushels could pass from lake to lake. A second enlargement (1873-83) increased the depth of water to twelve feet; and a third, in 1887, gave the canal a uniform depth of fourteen feet, admitting the passage of vessels with a carrying capacity of 75,000 to 80,000 bushels. When this depth shall prevail throughout the entire system of the St. Lawrence canals, vessels of 1,600 to 1,800 tons register will be able to bring full cargoes from the Upper Lakes to Montreal, and to cross the ocean if their owners see fit.† In the meantime the Montreal

* Kingsford's "Canadian Canals" (Toronto, 1865) contains an elaborate history of the Welland and the financial difficulties that attended its construction. The Imperial Government seem to have contributed some £55,555 towards it, while stock was taken in the enterprise by individuals in the United States for £69,625, and by English capitalists, £30,137. The first vessels to pass through the canal are said to have been the schooners *Ann and Jane* and *R. H. Boughton*, in November, 1829. On the 5th of July, 1841, during the first session of the United Parliament of Canada, Lord Sydenham announced that Her Majesty had confirmed the bill for transferring the Welland to the Provincial Government.

Mr. McLennan states that the first Canadian vessel to pass through the Welland was the propeller *Ireland*, Captain Patterson.

† The schooner *Niagara*, built by Muirs, of Port Dalhousie, was sent to Liverpool with 20,000 bushels of wheat about the year 1860. Captain Gaskin, of Kingston, built several sea-going vessels, one of which he took over to Liverpool himself and sold her there. But experience has proved that vessels suited to the navigation of the lakes will never be able to compete successfully with ocean steamships of 10,000 tons.

Board of Trade are memorializing the Government to have the Welland enlarged so that the largest vessels navigating the lakes may be able to tranship their cargoes at Kingston or Prescott as they now do at Buffalo; in other words, to locate the ship canal projected by the Deep Waterways Commission on Canadian territory instead of on the American side of the Niagara River.

The Rideau Canal, connecting Kingston with Ottawa, was undertaken as a military work by the Imperial Government at the instigation and under the personal superintendence of Colonel John By, of the Royal Engineers, from whom the obsolete Bytown derived its name. A stupendous undertaking it was considered at the time—$126\frac{3}{4}$ miles long, with forty-seven locks, 134 feet by 32 feet each. It was begun in September, 1826, and on the 29th of May, 1832, the works being completed, the steamer *Pumper* passed through from Bytown to Kingston. The limit of this canal is a draught of five feet. Its cost is said to have been about one million pounds sterling. It was transferred by the Imperial authorities to the Provincial Executive in 1856.

The St. Lawrence Canal System, with a uniform depth of nine feet of water, was completed in 1848. The canals are eight in number, viz.: the Lachine Canal, $8\frac{1}{2}$ miles; the Beauharnois, $11\frac{1}{4}$ miles; the Cornwall, $11\frac{1}{2}$ miles; Farren's Point, $\frac{3}{4}$ of a mile; Rapid du Plat, 4 miles; Galops, $7\frac{3}{8}$ miles; the Welland, $26\frac{3}{4}$ miles, and the Sault Ste. Marie, $\frac{3}{4}$ of a

CANADIAN SHIP CANAL AT SAULT STE. MARIE, 1895.

mile—in all 71¼ miles, with 53 locks, and 551¼ feet lockage. In 1871 the Government decided to enlarge the locks of the whole system to 270 feet by 45 feet, and to deepen the canals to fourteen feet. These dimensions were decided upon after consultation with the Boards of Trade of Oswego, Toledo, Detroit, Milwaukee and Chicago; but so great has been the increase of commerce on the lakes since then, so much larger are the vessels now employed in the trade, and so keen has the competition become in the transportation business, it is already apparent that the limiting of the locks to 270 feet has been a mistake, and that before the work in hand is finished there will be a call for locks of at least double that capacity.

Under the new arrangement the Lachine Canal has two distinct systems of locks, giving two entrances at each end. The Cornwall Canal has in the same manner two sets of locks at its lower entrance, and has in other respects been greatly improved. The Beauharnois Canal was not enlarged, but, instead, an entirely new canal on the north shore of the river is being constructed, fourteen miles in length, of the same dimensions as the other enlarged canals, at a cost of $5,000,000. The total cost of the St. Lawrence canals and river improvements west of Montreal has been $29,000,000; of the Welland Canal, $24,000,000; the Sault Ste. Marie, $3,258,025; of the Ottawa and Rideau canals, about $10,000,000; and of the whole canal system of the Dominion about $75,000,000. The total revenue derived from tolls and hydraulic and other rents for the year 1895 was $339,890.49; 2,412 vessels passed through the Welland during the season of 1894, carrying 1,008,221 tons of freight. The quantity of freight moved on the St. Lawrence and Ottawa canals was 1,448,788 tons, and on all the canals over 3,000,000 tons, whereof the products of the forest, 1,077,683 tons; agricultural products, 993,348 tons— the remainder being general merchandise and manufactures.*

The deepening of Lake St. Peter and other shallow reaches of the St. Lawrence between Montreal and Quebec has created what may be called a submerged canal, fifty miles long, three hundred feet wide, with a minimum depth of $27\frac{1}{2}$ feet, permitting ocean steamers of the largest class now in the trade to discharge their cargoes in the port of Montreal, which is undergoing enlargement at the present time at a cost of many millions of dollars.

During the season of 1897 the number of sea-going vessels that arrived at Montreal was 796, with a total tonnage of 1,379,002; 752 of these were steamers, aggregating 1,368,395 tons. The inland vessels numbered 6,384, with a tonnage of 1,134,346. The sea-going steamers were eighty-three in excess of the previous year, with a marked increase of tonnage.† During that summer steamships of 10,000 and even 12,000 tons burthen were to be found loading and discharging cargo alongside the wharves of Montreal.

* " Report of Dominion Railways and Canals, 1895," p. 256.
† " Montreal Board of Trade Report, 1897," p. 70.

The total value of merchandise exported from this port during the year 1897 was $55,156,956. The chief articles of export were as follows:

	Quantity.	Value.
Produce of the mines	......	$188,127 00
" " fisheries	......	120,242 00
" " forest	......	5,731,583 00
Horses (Number)	12,179	1,205,941 00
Horned Cattle "	119,188	7,151,280 00
Sheep "	66,319	340,060 00
Butter (Pounds)	10,594,824	1,878,515 00
Cheese "	162,322,426	14,325,176 00
Eggs (Dozen)	4,806,011	575,782 00
Meat of all kinds (Pounds)	16,377,806	1,345,894 00
Wheat (Bushels)	9,900,308	8,415,261 00
Indian Corn "	9,172,676	3,121,753 00
Other grains (barley, oats, peas, etc.) "	10,298,444	3,904,128 00
Flour (Barrels)	891,501	3,120,253 00
Apples "	175,194	350,000 00
Manufactured and miscellaneous articles	......	3,954,919 00

CHAPTER IX.

STEAM COMMERCE OF THE GREAT LAKES.

United States and Canadian Commerce of the Great Lakes—The Sault Ste. Marie Ship Canals—The Erie Canal—Transportation Business—The Elevator—Deeper Waterways Commissions—The Ottawa and Georgian Bay Canal.

DURING the last quarter of a century the commerce of the Great Lakes—the United States commerce especially—has grown with a rapidity almost exceeding belief. It has become enormous! At the present time it is stated on competent authority that the steam tonnage of these inland seas largely exceeds the combined tonnage of this character in all other parts of the United States put together. Not to speak of the vast amount of shipping employed in the iron, the coal, and the lumber trade, the Lake Superior grain and flour shipments for 1896 were 121,750,000 bushels. The Lake Michigan grain and flour shipments for the same year were 273,820,000 bushels, together making 395,570,000 bushels of grain and flour shipped in one year from these two quarters! It is difficult to realize the magnitude of such a statement. Mr. Keep, already quoted, in his report for

1890 puts it strikingly when he says: "If the freight carried on the Great Lakes in the United States coastwise and foreign trade during the year 1890 were loaded into railway cars of average size and capacity, the cars so loaded would cover 13,466 miles of railroad track." The Commissioners appointed by the Canadian Government to meet with a similar Committee appointed by the United States Government to consider the subject of international and deeper waterways, preface their report by alluding to the commerce of the Great Lakes in these terms: "It is impossible to convey, within reasonable space, an adequate idea of the extraordinary* development of inland water transportation on the Upper Lakes—which for rapidity, extent, economy and efficiency has no counterpart even on the ocean. More than half of the best steamships of the United States are imprisoned above Niagara Falls, and more than half of the tonnage built in the United States in 1896 was launched upon the lakes." This inland water commerce has built up twelve cities on the southern shores above Niagara, five of which have over 200,000 population, and one of them over a million. Within these limits there are twenty-seven dry docks, the largest of which is on Lake Superior and is 560 feet long, 50 feet wide, and 18 feet depth of water. There are sixty-three life-saving stations upon these lakes, ten of which are Canadian. "Unusual prosperity has stimulated ship-building to such an extent that there are now in course of construction at the various

---

* *Vide* page 26 of said Report.

lake shipyards, sixty-five vessels, thirty of which are steel freight steamers which will average 400 feet in length and 4,000 tons capacity—costing in all $9,000-000.*

Up to a comparatively recent date the bulk of the lakes commerce was done by sailing vessels. Every town of any importance had its little fleet of schooners. As time went on, the vessels increased in size, and eventually a very fine class of three-masted schooners, with some brigs, barquentines, and even full-rigged barques, were employed in the carrying trade. One of the largest of these was the barque *Utica*, of 550 tons, which sailed on the Buffalo and Chicago route in the forties. A few of these clipper schooners may still be met with, but they are rapidly being supplanted by iron and steel steamships of great size, such as the *Maryland*, the *Owego*, the *E. C. Pope*, and the *Manitou*, representatives of fleets of first-class steamships, ranging from 300 to 350 feet in length, over 1,900 tons register, with triple expansion engines, a speed of from fourteen to sixteen miles an hour, and a carrying capacity of 120,000 to 125,000 bushels of grain. These, and many others like them, were accounted "queens" a few years ago; they are fine ships still, but there are much larger and finer than they now.

The *Manitou* here represented is one of the finest ships of her class on the lakes, built in 1893 by the Chicago Ship-building Company. Her hull is of steel, length over all 295 feet, breadth of beam 42 feet,

---

* "Buffalo Board of Trade Report, 1895," p. 98.

and depth of hold 22 feet. Her average draught of water is 15 feet. She has triple expansion engines, a single four-bladed screw propeller 13 feet in diameter. Her gross tonnage is 2,944 tons. She is handsomely fitted up with sleeping accommodations for four hundred passengers, has a freight capacity of 1,500 tons, and develops a speed of eighteen miles an hour. Her route is between Chicago and Sault Ste.

THE "MANITOU," 1893.

Marie, where she connects with the Lake Superior lines. She cost $300,000.

The *James Watt*, the first of the Rockefeller fleet and the largest steamship on the lakes, is 426 feet long, 48 feet beam, and 29 feet deep. She cost $260,000, and will carry from 4,000 to 6,000 tons of ore, according as she is trimmed to draw 14 or 18 feet of water. The *Empire City*, owned by the

Zenith Transportation Company, is of the same dimensions, less one foot in depth. She is now the largest grain carrier on the lakes, having capacity for 213,000 bushels. The Minnesota Iron Company have a fleet of fourteen steamships, each carrying from 100,000 to 180,000 bushels of grain. The Lehigh Valley Transportation Company own a fleet of large and powerful steel freight steamers which ply between Buffalo and Chicago. These are but a few of the many transportation companies that do business on the Great Lakes. As to the vessels at present employed in the trade, it is safe to say that they are to be regarded only as the precursors of a still larger class of freight steamers that will navigate these waters when the contemplated twenty-one foot channel shall have been established from Lake Superior to Buffalo. At present there is a navigable channel of $17\frac{1}{2}$ feet all the way.

Many of the large steamers take a number of barges in tow, and in this way enormous quantities of grain are sometimes moved by a single shipment. The *Appomattox*, for example, with three consorts in tow, recently left Duluth with a combined cargo of 482,000 bushels, or 14,460 tons of wheat. Assuming the average yield of that cereal to be twenty bushels to the acre, this single shipment represented the produce of 24,100 acres!

The Northern Steamship Company of Buffalo has perhaps the finest fleet of steamers on the Great Lakes, consisting of eight steamships. Six of these are steel freight and emigrant ships of 2,500 tons each. They

SS. "NORTH-WEST," 1894.

are named the *Northern Light*, *Northern Wave*, *Northern King*, *Northern Queen*, the *North Star*, and the *North Wind*. The other two, the *North-West* and the *North-Land*, are exclusively passenger ships, up-to-date in every respect. They are identical in size, being each 386 feet long, 44 moulded breadth, and 26 feet in depth. Their gross tonnage is 5,000 tons apiece. They have quadruple expansion engines of 7,000 indicated horse-power. The boilers are worked at a pressure of 275 pounds to the square inch, and use up 70 tons of water per hour. The twin screws are 13 feet in diameter and 18 feet pitch, make 120 revolutions per minute, and drive the ships at a speed of from 22 to 25 statute miles an hour, as may be required. The bunkers hold 1,000 tons of coal. A double bottom, 42 inches deep, extends the whole length of the ship, and is utilized for adjustable water ballast. Luxurious accommodation is provided for five hundred first-class and forty second-class passengers. Nearly twenty-six miles of electric wire are used in conducting the subtle fluid for 1,200 lights. The electric search-light has one hundred thousand candle-power. The refrigerating plant, besides creating ample cold storage, makes one thousand pounds of ice per day for the ship's use. The grand saloon is, in American parlance, "a magnificent achievement." The routes of these twin ships is from Buffalo to Duluth, at the head of Lake Superior, a distance of 1,065 miles, each of them making the round trip in a week. The fare for the round trip is $30 for transportation, meals and staterooms being charged extra.

For many years two causes prevented the building of vessels of such large dimensions as those just described for lake navigation. One of these was the insufficient size of the lock at Sault Ste. Marie, and the other was the shallowness of the water on the St. Clair flats and at other points. The former difficulty disappeared in 1881 when the first of the large locks was opened at the Sault; the second difficulty was overcome by the Northern Steamship Company in the peculiar construction of their vessels with a water ballasting system that permits of sinking the ship to the depth required for navigating the deep waters of the lakes and of floating them over the shoals and bars that obstruct the navigation. This ingenious device, however, can only be regarded as a temporary expedient, pending the action of the United States Government, which contemplates the making of a twenty-one foot channel at all points where the shallows occur. This is a measure felt to be due to the lakes' marine, which has already done so much to develop the resources of the North-West, especially the mineral resources, which would otherwise have lain comparatively dormant. "The United States have expended some $12,000,000 in widening and deepening channels, which has already been more than repaid by the rapid development of commerce. The largest item in the lakes' traffic is the transportation of iron—the richest ores are now being mined along a line of coast of one thousand miles, dotted with manufacturing towns."[*]

[*] "United States Deep Waterways Commission Report, 1896."

It helps one to realize the immensity of the lakes' traffic to learn that the number of vessels that cleared from the district of Chicago in 1893 was 8,789, with a gross tonnage of 5,449,470 tons—actually a larger tonnage than cleared from the port of Liverpool in 1892.* The tonnage passing down the Detroit River from lakes Superior, Michigan and Huron, during the seven or eight months of navigation, is, by official statements, greater than the entire foreign and coastwise trade of London and Liverpool combined in twelve months. It is estimated by competent experts to be three times greater than the foreign trade of the port of New York, and to exceed the aggregate foreign trade of all the seaports of the United States by 10,000,000 tons!

Sault Ste. Marie Ship Canals.

To accommodate the vast volume of traffic emanating from Lake Superior ports, magnificent canals have been constructed on either side of the St. Mary River, which connects Lake Superior with Lake Huron. These works, the most remarkable of their kind in existence, have reached their present dimension by a succession of enlargements and a large outlay of money. The first canal on the western or American side of the river was constructed by a joint stock company formed in 1853, who undertook to construct it for the State of Michigan upon receiving therefor a grant of 750,000 acres of land. The work

---

* "Chicago Board of Trade Report, 1895."

was completed in 1855, and from that date the commerce of Lake Superior may be said to have had any appreciable existence. The opening of the canal was, as it were, the opening of a sluice-gate through which a flood of commerce was soon to roll.

The first canal cost about $1,000,000. It was a little over a mile in length. Its width at the water line was 100 feet, and its depth 12 feet. There were two locks, each 350 feet long and 70 feet wide. The growth of traffic and the increase in the size of the lake vessels soon rendered it apparent that the canal must be enlarged. In 1870 the United States Government made its first appropriation for deepening the canal to 16 feet and increasing its lockage. A new lock was built, 550 feet in length by 80 feet in width, and 18 feet lift, at a cost of $2,404,124.33. The work was completed in 1881. Its opening was followed by an enormous increase of commerce—so much so that it soon became quite inadequate to the traffic. A still further enlargement was decided upon, and was completed in 1896, at a cost of about $5,000,000. The new lock occupies the site of the two old locks of 1855, and is 800 feet long, 100 feet wide, and has 21 feet depth of water on the sill. It is officially known as the St. Mary's Falls Canal.

So long ago as the close of last century the North-West Fur Company had constructed a rude canal on the Canadian side, with locks, adapted for the passage of loaded canoes without breaking bulk. Though late of construction, a ship canal had long been in contemplation by the Canadian Government, and the time came when, owing to the increase of traffic, it could no longer be delayed. This great work was completed and opened for traffic on September 9th, 1895, at a cost of some $3,500,000. The Canadian lock is 900 feet long, 60 feet wide, 20 feet 3 inches depth of water on the sill, and 18 feet lift, affording room for three large vessels at one time. The length of the canal proper, between the extreme ends of the entrance piers, is only 5,967 feet, but including the excavated channels of approach it is about 18,100 feet. The American canal is a little over a mile in length. The locks of both are unsurpassed for their size and solidity, as well as for the completeness of their mechanical apliances.

An official report, compiled by the Chief Engineer of the St. Mary's Falls Canal (United States), contains a detailed statement of the commerce of that canal for each year, from 1855 to 1895, and goes far to substantiate what has already been said as to the magnitude of the lakes' commerce. The number of vessels that passed through in 1895 was 17,956, with a registered tonnage of 16,806,781 tons. The number of sailing vessels was 4,790; of steamers, 12,495; and of unregistered craft, 671. The number of passengers conveyed from lake to lake was 31,656. As to the cargoes of the vessels, these are a few of the chief items: 2,574,362 net tons of coal; 8,902,302 barrels of flour; 46,218,250 bushels of wheat; 8,328,694 bushels of other grain; 107,452 tons of copper; 8,062,209 tons of iron ore; 740,700,000 feet

ST. MARY'S FALLS SHIP CANAL OF 1881, STATE OF MICHIGAN, U.S.A.

of sawed lumber; 100,337 tons of manufactured and pig iron; 269,919 barrels of salt—in all, 15,062,580 net tons of freight. The freight traffic of the St. Mary's Canal, in seven months of 1895, was more than twice that of the Suez Canal, which is open all the year. During the year 1897 it was much greater than in any previous year, the registered tonnage being 17,619,933, the tons of freight 18,-218,411, and the number of passengers 40,213.

The gradual development of steam navigation on Lake Superior is shown in a table of parallel columns, extending over thirty years. In 1864 the sailing vessels that used the canal were three times as numerous as the steamers; but in 1895 the steamers were three times as numerous as the sailing vessels, and they had increased enormously in tonnage. The number of sailing vessels built on the Great Lakes in 1896 was nineteen; in that year there were built *seventy-five* steamers, aggregating 75,743 tons register, and of these thirty-five were built of steel, with a combined tonnage of 63,589 tons. The principal ship-building yards on the Upper Lakes are at Buffalo, Cleveland, Detroit, Bay City, Milwaukee, Chicago and Superior. At most of these points there are plants for the construction of iron and steel vessels. It is said that Cleveland is the largest ship-building port, and also the largest iron ore market in the United States.

The transportation of iron ore, it will be noticed, forms a large element in the commerce of Lake

Superior. Not only is the ore found in great abundance in that region, but it is the best in quality and the most in demand of any in the United States. Over 100,000,000 tons of this ore have been mined in the lake region within the last forty years. Owing to its great bulk and weight it is nearly all carried by water; the estimated capital engaged in mining and transporting the ore to the 120 furnaces in Ohio, Pennsylvania, Buffalo and Chicago is about $234,000,000.* But for the number and the size of the steamers thus employed, and the facilities now in use for loading and unloading them, the trade could not exist. The largest vessels in the iron ore trade are regularly loaded in three or four hours; 2,500 tons of ore have been loaded into a vessel of that capacity in an hour and three-quarters. †

The Erie Canal.

This great artificial waterway, lying wholly in the State of New York, and under State management, connects Buffalo with the Hudson River at Albany. Although of comparatively limited capacity, it is to-day the most formidable rival the St. Lawrence route has to compete with in the transportation of freight from the west to the seaboard. The Erie was first opened for traffic in the same year as the first Lachine Canal (1825). It was originally 363

---

*" United States Deep Waterways Commission Report, 1896."
†" United States International Commerce Report, 1892," p. 52.

miles long, with eighty-three locks, each 90 feet by 15 feet, and 4 feet depth of water.

The first enlargement of this canal was commenced in 1836 and completed in 1862, at a cost of $44,465,414, making the entire cost up to the last-named date over $50,000,000. It is now $351\frac{3}{4}$ miles in length, 70 feet wide on the surface and 56 feet wide at the bottom having 72 locks, each 110 feet by 18 feet, and 7 feet deep. The limit of the canal for navigation, however, is only 6 feet of water, restricting its use to vessels of 240 tons capacity, say, 8,000 bushels of wheat.

Navigation has hitherto been carried on by horse traction—the boats running in pairs—and by small steam tugs towing three or four boats, after them. The tug often pushes one boat ahead and tows the others behind. In this latter way a load of 900 tons will be moved at an average pace of about $2\frac{1}{2}$ miles per hour while in motion. Including lockages, the distance from Buffalo to New York may be covered in nine or ten days. The boats are about 98 feet long and 17 feet 5 inches wide. They make on the average about seven round trips in the season. The average price received for the transportation of wheat in this manner from Buffalo to New York is about $3\frac{1}{2}$ cents per bushel, which allows a fair margin of profit to the boatman.

Experiments have been made for the application of electricity to the traction of the boats, with promise of further development. In the meantime considerable importance is attached to the installation of electric telephone communication from one end of the line to the other, whereby instant communication can be had with the section superintendents, the lock tenders and other officials. The system is devised solely for the use of the canal officials, and will be invaluable in sudden emergencies caused by accidents to the boats, leaks, breaks, or other disasters that may occur and interfere with the navigation of the canal.

For some time past western shippers have been testing the feasibility of establishing a through line of transportation from the Great Lakes to New York by way of the Erie Canal without the delay and expense of transhipment at Buffalo. In 1895 this idea was worked out by the construction of a fleet of steel canal boats, consisting of one steamer and five consorts, by the Cleveland Steel Canal Boat Company of Ohio. Several fleets of this kind have since been put in operation, and the projectors believe that they have demonstrated the practicability of thus carrying freight to the seaboard from any of the western lakes at a fair margin of profit and in successful competition with the railways. These steel barges have encountered severe storms on the lakes without any serious damage to the boats or their cargoes. The cost of the tug boat is about $15,000, and of each consort about $6,000. The time occupied by the steel fleet from Cleveland to New York has been from ten to twelve days.

The second enlargement of the Erie Canal, now in progress and nearing completion, will afford greatly increased facilities for transportation, by increasing the depth from 7 to 9 feet and doubling and lengthening all the locks. There will be no increase in the width of the locks nor in the length of the boats navigating the canal, but two boats (which form a horse-tow) will be locked through at once, and by the locks being doubled, side by side, no boats will have to wait for others coming in an opposite direction. The cargo will be increased by the greater depth of water in boats of the same size, more deeply loaded, and the traction will be so improved that boats will run easier and faster. The amount of freight carried on the Erie Canal—east and west—in the year 1896 was 2,742,438 tons.* The amount transported on the Welland Canal for that year was 1,279,987 tons.

Canadian Commerce on the Great Lakes.

Notwithstanding the large amount of money expended by the Canadian Government upon its unrivalled St. Lawrence canals and the deepening of its waterways, the volume of western traffic that comes this way is as yet disappointingly small. The great bulk of the trade in western produce, Canadian and American, finds its way to the seaboard in American vessels by way of Buffalo, Oswego and Ogdensburg to New York and Boston. What effect the deepening of

* For these notes on the Erie Canal the author is chiefly indebted to Kingsford's "Canadian Canals," Mr. Thomas C. Keefer, C.E., Ottawa, and the Superintendent's "Report on Canals in the State of New York, 1896."

the canals to fourteen feet will have on this deviation from the "natural outlet" remains to be seen.

From a statement kindly furnished by Mr. T. F. Taylor, Marine Inspector at Kingston, it appears that the number of companies in Canada having steamers and other craft engaged in the commerce of the Great Lakes is twenty-four. Three of these go no farther than the head of Lake Ontario, three extend their operations to Lake Erie, five to Lake Huron, and thirteen to Lake Superior. Five steamers are employed on Lake Erie, thirteen on Lake Huron, twenty-six navigate the waters of Lake Superior. About one-half of these steamers are first-class steel freight and passenger vessels of from 1,200 to 2,600 tons each. A few of them pass through the Welland Canal and have their cargoes transhipped into barges at Kingston or Prescott. Others connect with lines of railway at Sault Ste. Marie, Owen Sound, Collingwood, Windsor and Sarnia. Occasionally one or two of the smaller ones run through to Montreal. Besides the steamers, there are employed in the lakes' grain trade twenty-one lake barges, each of 50,000 bushels capacity, and fourteen tug steamers. There is also a fleet of about sixty-two sailing vessels trading between the Upper Lakes and Kingston, and some sixty or seventy barges employed in transporting grain from that port to Montreal.

On the completion of the Canadian Pacific Railway the company formed a line of freight and passenger steamers of their own, consisting of the *Algoma*, the *Alberta* and the *Athabasca*. The *Algoma* had sailed

C. P. R. SS. "ALBERTA," 1883.

the lakes previous to this under different names. The other two are fine steel ships, built by Aitken & Co. of Glasgow, in 1883. They are each 270 feet long and 2,300 tons burthen, fitted with all modern improvements in their machinery and with excellent accommodation for a large number of passengers. They commenced their work in 1884 and have been very successful and popular. The *Algoma* was unfortunately wrecked off Isle Royale in Lake Superior in November, 1885, during a fearful snow-storm that swept over the lake, when many lives were lost. She was replaced by the *Manitoba*, a very fine vessel built of steel at Owen Sound by the Polson Ship-building Company. The *Manitoba* is the largest Canadian steamer on the lakes, being 300 feet long and 2,600 tons burthen. By means of these steamers a regular and most satisfactory summer service is maintained once a week from Windsor and Sarnia, and twice a week from Owen Sound and Sault Ste. Marie to Fort William. Their capacity for the transportation of grain is about 400,000 bushels a month.

The Montreal Transportation Company, founded in 1868, is the oldest of the existing forwarding companies, and does the largest amount of business. Their fleet consists at present of three steamers, six tugboats, six lake barges and thirty-two river barges. Two of the steamers, the *Bannockburn* and the *Rosemount*, are first-class steel ships, built at Newcastle-on-Tyne, about 250 feet in length, 40 feet beam, with a carrying capacity of 75,000 bushels of wheat. The lake barges play an important part as "consorts" to the steamers. They resemble in appearance so many large dismasted schooners, and serve their purpose economically and well so long as they keep in tow, but when they break loose, as they occasionally do when overtaken by a gale of wind, they become unmanageable and are apt to come to grief. This company with its present equipment handles about 250,000 bushels of grain per month.

The North-West Transportation Company, dating from 1871, and otherwise known as the "Beatty Line," has two fine passenger and freight steamers, the *Monarch* and the *United Empire*, of 1,600 tons and 1,400 tons respectively, forming a weekly line from Windsor and Sarnia to Fort William and Duluth, in connection with the Grand Trunk Railway; they forward about 200,000 bushels of grain per month.

The Hagarty and Crangle Line, running between ports at the head of lakes Superior and Michigan to ports on the River St. Lawrence, has two large steel steamers, the *Algonquin* and the *Rosedale*, on the Upper Lakes, and the steamer *Persia* which plies between the head of Lake Ontario and Montreal. Hamilton has three "Merchants Lines" in the Upper Lakes' shipping business—Mackay's, Fairgreaves', and Thomas Myles & Sons, owning in addition to other lake craft such fine steel and composite steamers as the *Sir L. Tilley*, *Lake Michigan*, *Arabian* and the *Myles*.

The Calvin Company's Line, of Garden Island, Kingston, has four steamers, four lake barges, and four tug steamers running between Lake Superior ports, Kingston and Montreal. The Collins Bay Rafting Company has on the same route three steamers, three lake barges, and two tug steamers. The Jacques & Co.'s Line has two steamers running from the head of Lake Erie and one from the head of Lake Ontario to Montreal.

The Great Northern Transit Company, with headquarters at Collingwood, has four freight and passenger steamers—the *Majestic, Pacific, Atlantic,* and *Northern Belle*—keeping up a well-appointed service twice a week from Collingwood to Sault Ste. Marie, and having connection with the Northern Railway to Toronto. The *Majestic*, built at Collingwood, is a steel screw steamer, 230 feet long, 36 feet wide, 1,600 tons register, and cost $125,000. She has compound condensing engines of 1,200 horse-power, and is fitted up internally with great elegance. The North Shore Navigation Company has five excellent steamers plying on the Georgian Bay and northern shores of Lake Huron from Collingwood and Owen Sound to Sault Ste. Marie and Mackinac Island, where connections are made with American lines of steamers to Chicago and other ports on Lake Michigan. The steamers are the *City of Collingwood*, 1,400 tons; *City of Midland*, 1,300 tons; *City of Toronto*, 800 tons; *City of Parry Sound* and *City of London*, each 600 tons.

Reference will be made hereafter to steamers plying on Lake Ontario and the River St. Lawrence.

The Transportation Business.

In the matter of transportation it may be interesting to learn how a consignment of wheat is "handled" from the time it leaves the field in Manitoba, where it is grown, until it reaches its destination in Liverpool or London. When there

C. P. R. GRAIN ELEVATOR AT FORT WILLIAM, ONT.

were only a few hundred thousand bushels to be sent to the seaboard, the means of transport were very simple and primitive. It was carried on men's backs from one conveyance to another, and floated down rivers or shallow canals in small boats or on rafts of timber. But when the thousands became millions the problem of cheap transportation became a serious one. American ingenuity rose to the occasion and invented the most marvellous of labour-saving appliances—THE GRAIN ELEVATOR.

The farmer sells his crop of wheat to the grain-dealer, and carts it, say, to Brandon, where the purchaser takes delivery of it at his elevator. Let us examine this thing somewhat minutely, taking by way of illustration one of the elevators belonging to the Canadian Pacific Railway Company at Montreal. It is a medium-sized one, having capacity for storing about 600,000 bushels of grain. The same company's elevators at Fort William and Port Arthur are much larger, having capacity for 1,500,000 bushels. In Chicago and Buffalo there are elevators of three millions of bushels capacity; but, whether larger or smaller, in their main features they are all alike.

The elevator is a wooden structure of great strength. Its massive stone foundations rest on piles imbedded in concrete. The framework is so thoroughly braced and bolted together as to give it the rigidity of a solid cube, enabling it to resist the enormous pressure to which it is subjected when filled with 18,000 tons of wheat. The building is 210 feet long, 80 feet wide, and 142 feet in height from basement to the peak of the roof. Including the steam-engine (built at the C. P. R. works) of 240 horse-power, the entire cost of this elevator was $150,000. It consists of three distinct compartments—for receiving, storing, and delivering grain. On the ground floor are two lines of rails by which the cars have ingress and egress. The general appearance of this flat is that of a bewildering array of ponderous posts and beams, shafting, cog-wheels, pulleys and belts, blocks and tackle, chutes, and the windlasses for hauling in and out the cars, for a locomotive with its dangerous sparks may not cross the threshold. Beneath this, in the basement, are the receiving tanks, thirty-five feet apart from centre to centre, corresponding to the length of the cars. Of these there are nine, enabling that number of cars to be simultaneously unloaded. This is quickly done by a shovel worked by machinery, with the aid of two men, the grain falling through an iron grating in the floor into the tank. The elevator has nine "legs." The leg is an upright box, 12 inches by 24 inches, extending from the bottom of the tank to the top of the building; inside of it is a revolving belt with buckets attached $15\frac{1}{2}$ inches apart. The belt is 256 feet long, and as it makes 36 revolutions per minute, each bucket containing one-third of a bushel, each leg is able to raise 5,250 bushels per hour.* A car is unloaded and its contents hoisted

---

* The latest improvement in this direction is what is called the "Grain Sucker," by which the process of loading and unloading cargoes of grain is accomplished with astonishing speed. The new appliance combines in its construction the main features of the ordinary elevator, and causes the grain to go through all the different movements above described, with this difference, that instead of the leg with the belt and bucket, the grain is elevated to the top of the structure on the principle of suction through a flexible pipe. The air being drawn off by pumps from the vacuum chamber, the grain is sucked up like water from a well. Machines of this kind, fitted with any number of these pipes that may be required, are used at the London docks, and are said to be capable of transferring wheat at the rate of a hundred and fifty tons an hour—*Vide Strand Magazine* for **May, 1898.**

into the upper regions in fifteen minutes. When all the legs are at work 30,000 bushels are handled in an hour.

The four-story house on top of the granary contains a number of different mechanisms. In the uppermost flat the leg's revolving belt turns round a pulley and discharges the grain into a receiving hopper on the next floor. From this it is withdrawn to the weighing hopper, nicely balanced on a Fairbanks beam-scale, having a capacity for 30,000 pounds or 500 bushels of wheat, which is weighed with as much exactitude as is a pound of tea by the grocer. At either end of this room there is a separating machine in which the grain can be thoroughly cleansed by the removal of smut and dust. Underneath is the distributing room, with jointed pipes leading to the storage bins, of which there are one hundred, each 50 feet deep and 12 feet square, calculated to hold 6,000 bushels each. The process of withdrawing the grain from the bins, strange to say, is a repetition of that just described. It must go down into the cellar, and up again to the attic, and pass through the weighing machine and thence to the car, the barge, or the ship. A car of 600 bushels can be loaded in three minutes. The most singular part of the whole apparatus is the "carrier" by which the grain is conveyed from the elevator to the vessel lying at the wharf, 200 feet off. The carrier is an endless four-ply rubber belt, 515 feet long and 36 inches wide, upon which the grain is dropped and carried to its destination. The difficulty of comprehending why the grain is not shaken off that flat, rapidly revolving belt is not lessened by the explanation given, that it is held in place by the concentrative attraction of the particles in motion. But from whatever cause, the grain clings to the belt, and may be carried in this way any distance, and in all manner of directions, turning sharp corners and even going over the roofs of houses if they stand in the way. The elevator in question delivers by "carrier" from 8,000 to 10,000 bushels an hour. There are over 50 such elevators in New York, only of much larger capacity; Buffalo has 52, with a storage capacity of over 15,000,000 bushels; Chicago, 21; Duluth and Superior, 9 each. There are elevators in Buffalo that can take grain out of a vessel at the rate of 25,000 bushels an hour.

A Duluth paper of May 21st, 1898, says: "Globe elevator No. 1 carries the broom for rapid loading this year, and the record made yesterday has probably never been equalled. The steamer *Queen City* loaded there yesterday morning, taking 185,000 bushels in 180 minutes."

Now, suppose that an order has reached Brandon for a shipment of 220,000 bushels of wheat,* to be forwarded to Montreal *via* the St. Lawrence route. The initial cost of receiving, storing for a given time and delivery from the Brandon elevator is three cents per bushel. It must be hauled from Brandon to Fort

* "The steamship *Bannockburn* and consorts left Fort William on the 3rd instant loaded with 220,000 bushels of No. 1 hard wheat for Mr. W. W. Ogilvie's mills. This is the largest shipment that ever left the port."—*Montreal Gazette, June 5th, 1896.*

William, a distance of 559 miles by railway. The consignment is the produce of 11,000 acres and weighs 6,600 tons. It will load 330 box-cars, each containing 40,000 pounds. As each car weighs about 25,000 pounds, the entire weight to be moved by rail will be 10,725 tons. Until quite recently, twenty cars of wheat made up an average train load, but with the powerful locomotives now in use twice that number may be taken at a load. A safe estimate for this particular shipment will be ten trains of thirty-three cars each, the gross weight of engine, tender and train being about 1,100 tons.* The cost of transport from Brandon to Fort William, at the summer rate of 19 cents per 100 pounds, will be 11.40 cents per bushel. By means of the elevator at Fort William it is transhipped to lake vessels. A large propeller takes on board 70,000 bushels; the balance is stored away in three barges containing 50,000 bushels each. The propeller takes the trio in tow and proceeds on its long voyage of 1,200 miles through Lake Superior, the "Soo" Canal, lakes Huron and Erie, the Welland Canal and Lake Ontario to Kingston, in seven days.

* The weight that can be hauled by a locomotive depends largely on the gradients of the road traversed. Winnipeg and Fort William are nearly on the same sea level, but between them the line of railway ascends and descends some 800 feet, limiting the drawing power of a sixty-ton locomotive in certain sections to, say, 900 tons. On a level road a large American locomotive will easily draw sixty cars containing 1,000 bushels of wheat each, or a total weight of 3,000 tons. As with steamships, the tendency is to increase the size of the locomotive. There is this difference, however: the weight and power of the locomotive are limited by the strength of the rail upon which it travels.

The cost of transportation from Fort William to Kingston is from three to four cents per bushel, and to Montreal two cents more. At Kingston floating elevators come alongside the propeller and her consorts, and quickly transfer their cargoes into lighters carrying from 20,000 to 30,000 bushels each.* The fleet of nine or ten river barges is towed down the St. Lawrence, passing through the Cornwall, Beauharnois and Lachine canals to Montreal, 1,940 miles from Brandon by this route. They are laid alongside the ocean steamers in pairs, one opposite the forehatch and the other at the afterhatch, and their contents are poured into the big ship at the rate of 8,000 to 10,000 bushels per hour. The average rate to Liverpool is about $5\frac{1}{4}$ cents per bushel, bringing up the total cost of transportation from Brandon to Britain to, say, $22\frac{1}{4}$ cents per bushel. The first shipment of wheat from Manitoba to Britain was made in October, 1877.

* Since these lines were written, three stationary elevators have been erected at Kingston—one by the Montreal Transportation Company, with a capacity of 800,000 bushels; one by the Moore Company, for 500,000 bushels, and one by James Richardson & Sons, for 250,000 bushels. The Prescott Elevator Company has erected one at Prescott of 1,000,000 capacity, and still another has been built at Coteau Landing in connection with the Canada Atlantic Railway system, with 500,000 capacity. All indications are that the enlargement of the St. Lawrence canals is confidently expected to result in a large increase in the Canadian grain trade and forwarding business. There are sixteen floating elevators in Montreal harbour, capable of handling from 4,000 to 8,000 bushels of grain each per hour.

Mr. Hugh McLennan, the president of the Montreal Transportation Company, is also one of the most extensive shippers of grain in Canada. No better illustration can be found anywhere of the man who is the architect of his own fortune. Mr. McLennan was born in the County of Glengarry in 1825. His father's family came from Ross-shire, Scotland, in 1802, and his mother's family were United Empire Loyalists, who settled in Glengarry at the close of the American War of Independence.

After serving some years in the hardware business in Montreal, Mr. McLennan joined the mail steamer *Canada*, as purser, under Captain Lawless. In 1850 he started business on his own account in Kingston, as wharfinger and shipping agent. During that season he united with some others in organizing a steamboat line to run between Kingston and Montreal, in the furtherance of which enterprise he removed to Montreal in 1851, adding the business of general shipping agent. In the year 1854 he was joined by his elder brother John, when they entered extensively into the grain trade, Mr. McLennan going to Chicago in connection with that business. In 1867 he returned to Montreal, and organized the Montreal Transportation Company, of which he has been president to the present time.

Mr. McLennan's name soon became identified with many of the leading enterprises of the city, as well as in its educational and benevolent institutions. He still continues his active connection with the transportation and grain export business, and by reason of

his long connection has become an acknowledged authority in everything pertaining to the past history of these important branches of Canadian trade. He is an ex-president of the Board of Trade, and represented that organization upon the Harbour Board for a quarter of a century, resigning the position during the present season. He is a director of the Bank of Montreal, a governor of McGill University, and of the Montreal General Hospital, and is treasurer of the Sailors' Institute. He is also an active member of the American Presbyterian Church.

A large proportion of the wheat grown in the Western States and in Canada is made into flour and transported in that form to eastern and foreign markets. Minneapolis, in the State of Minnesota, claims to be the greatest flour manufacturing centre in the world Its milling capacity is said to be 54,800 barrels daily. Its actual output in 1895 was 10,581,-633 barrels. Although Canada may not compare with Minneapolis in its annual output of flour, she claims to have the largest individual miller in the world, in the person of W. W. Ogilvie, of Montreal. Mr. William Watson Ogilvie was born at St. Michel, near Montreal, April 14th, 1836, being descended from a younger brother of the Earl of Angus, who, some centuries ago, was rewarded with the lands of Ogilvie, in Banffshire, and assumed the name of the estate. His immediate ancestors belonged to Stirlingshire, Scotland, his grandfather having come to this country in the year 1800.

The milling business now represented by Mr. Ogilvie was begun by his grandfather, who, in 1801, erected a mill at Jacques Cartier, near Quebec, and also one at the Lachine Rapids, in 1808. In 1860 he became a member of the firm of A. W. Ogilvie & Co., then formed, whose transactions in grain soon became very extensive, resulting in the building of the "Glenora Mills," at Montreal, and others of large capacity at Goderich, Seaforth and Winnipeg. On the death of Mr. John Ogilvie, in 1888, Senator A. W. Ogilvie, having retired in 1874, Mr. W. W. became the sole member of the firm, and has since proved himself a man of marvellous executive ability. He went to Hungary to see the roller process at work, where it came into use in 1868, and was one of the first to introduce it into this country. He acquired by purchase the famous Gould Mills in Montreal, at a cost of $250,000, thus adding 1,100 barrels to his daily milling capacity, which, at the present time, is about 9,000 barrels a day. The annual output of Mr. Ogilvie's mills is about 2,500,000 barrels. About 30 per cent. of that amount is exported to different European countries; and, recently, a demand has arisen in Japan, Australia, and even in the Fiji Islands, for "Ogilvie's Hungarian flour." The balance is sold in all parts of the Dominion. Mr. Ogilvie purchases between four and five millions of bushels of wheat annually, and is rich in elevators, having as many as sixty-nine of these for his own special use in various parts of the country. In carrying on his extensive business he occasionally charters whole fleets of lake steamers and barges, and it is said of him that he is as fair in his business methods as he is generous in his charities. Mr. Ogilvie is a director of the Bank of Montreal, ex-President of the Montreal Board of Trade, and largely interested in several of the leading commercial interests of Canada.

Deeper Waterways.

The enlargement of the St. Lawrence and Erie canals cannot fail to prove advantageous to the inland shipping trade; but, so far from solving the question of " cheapest transportation," it seems rather to have accentuated the demand for greater facilities of a like kind. The cry for " deeper waterways " has been in the air for many years, but never has it been louder than just now. The first enlargement of the St. Mary's Falls Canal in 1881, and the subsequent deepening of the channels connecting the Upper Lakes had the effect, almost immediately, of doubling the tonnage of vessels plying the lakes and of producing a corresponding reduction in the rates of freight. The increase of the commerce of the lakes, incredible to those who are not engaged in it, and what appears to be its limitless future, have been keenly discussed in conventions as well as on the floors of Parliament and Congress for a number of years past, but it was only in 1894 that the movement assumed an organized form.

At a meeting held in Toronto in September, 1894, there was formed "The International Deep Waterways Association," the declared object of which was "to promote the union of the lakes and the high seas by waterways of the greatest practicable capacity and usefulness; and recognizing the supreme utility of such waterways' development." At that meeting it was resolved: "That the depth of all channels through the lakes and their seaboard connections be not less than twenty-one feet, and that all permanent structures be designed on a basis not less than twenty-six feet, in order that the greater depth may be quickly and cheaply obtained whenever demanded by the future necessities of commerce."

On the 8th of February, 1895, it was resolved by the Senate and House of Representatives of the United States of America in Congress assembled, "That the President is authorized to appoint three persons who shall have power to meet and confer with any similar committee which may be appointed by the Government of Great Britain or the Dominion of Canada, and who shall make inquiry and report whether it is feasible to build such canals as shall enable vessels engaged in ocean commerce to pass to and fro between the Great Lakes and the Atlantic Ocean, where such canals can be most conveniently located, and the probable cost of the same, with estimates in detail; and if any part of the same should be built on the territory of Canada, what regulations or treaty arrangements will be necessary between the United States and Great Britain to preserve the free use of such canals to the people of this country at all times."

By an order of Council dated at Ottawa, 14th December, 1895, Messrs. O. A. Howland, M.P.P., of Toronto, Thomas C. Keefer, C.E., and Thomas Munro, C.E., of Ottawa, were appointed Commissioners on behalf of the Canadian Government to meet and confer with the Commissioners appointed by the President of the United States on this important subject.

Several meetings of this International Waterways Commission have been held, a good deal of money has been spent in preliminary surveys, and reports favourable to the proposal, embodying much exceedingly interesting information as to the amount and rapid growth of the commerce of the lakes, have been submitted to the respective Governments. The American Commissioners favour the construction of a series of ship canals connecting Lake Erie with the seaboard, suggesting that the minimum depth of navigable water should be 28 feet, with canal locks 560 feet long and 64 feet wide. They present a choice of routes: (1) "The natural route" *via* the St. Lawrence to Montreal, and *via* Lake Champlain to the Hudson River. (2) *Via* Lake Ontario to Oswego and thence through the Mohawk Valley to Troy on the Hudson. The latter would be entirely through United States territory; the former would necessarily be of an international character, and preferable, provided that satisfactory treaty arrangements could be effected for the settlement of any differences that might arise between the two Governments interested. In either case the construction of a ship canal at Niagara Falls on the American side of the river is judged to be necessary. The international route would involve a ship canal from some point below Ogdensburg to near the boundary line on Lake St. Francis, and thence through Canadian territory to Lake Champlain.

The Canadian Commissioners in general terms endorse the international proposal as the one "which would give an opportunity of doing what our canals were intended to do, but have failed to do, that is, to obtain the maximum amount of the western trade for the St. Lawrence route." It is agreed that the class of vessels adapted to the Welland and St. Lawrence canals, limited to a draft of fourteen feet, can never compete successfully with the large United States vessels plying on the Upper Lakes; and the fact that these large United States vessels are unable to leave the Upper Lakes, "seems to embrace the whole 'Deep Waterways' question in a nutshell."*

Regarding Montreal as a seaport and the natural outlet for the commerce of the West, it is conceded

---

*The following paragraph, taken from the *North-Western Miller* for November 12th, 1897, doubtless reflects the opinion of the majority of Western grain dealers in the United States, with whom the feeling of sentiment for the "natural route" is of small account : "The steel barge *Amazon* left Fort William recently loaded with 205,000 bushels of Manitoba hard wheat for Buffalo, indicating that the Buffalo route is still at its best, and that the monster craft is cutting off the Montreal route as effectively as could be desired by any rival."

that its harbour accommodation must be largely increased, that it should be furnished with the best known appliances for the storage and shipment of grain, and that the navigable channel to Quebec be deepened to at least thirty feet and the Welland Canal to at least twenty feet.

The project of enlarged ship canals to connect the Great Lakes with the Atlantic Ocean is a magnificent one. Its advantages have been skilfully set forth. There are no insurmountable engineering difficulties in the way of its accomplishment, but it is still *in nubibus*. Apart from the complications necessarily attending an international work of this magnitude, it is not likely that it will be entered upon until the results to commerce of the enlargement of existing canals in both countries have been fairly tested.

In estimating the comparative merits of Montreal and New York, or any other American port on the North Atlantic coast, it may be assumed that the average summer rate of freight upon a bushel of wheat by either route from the head of the Upper Lakes to Liverpool is almost identical.* But it must

---

* We have good authority for quoting the rates of the summer of 1897 as follows : Duluth to Buffalo, 1½ cents per bushel ; Buffalo to New York, by the Erie Canal, 3½ cents ; New York to Liverpool, 5 cents ; elevator charges, ⅞ of 1 cent ; total, 10⅞ cents per bushel. Fort William to Kingston, 3½ cents ; Kingston to Montreal, 2 cents ; Montreal to Liverpool, 5¼ cents, including port charges ; total, 10¾ cents per bushel. In 1857 the average rate by lake and canal on a bushel of wheat from Chicago to New York was 25.29 cents per bushel ; now it is less than 6 cents. The reduction in cost of transmission is due to improved methods of handling freight, deeper channels, larger vessels and more rapid conveyance.

be borne in mind that grain shipped *via* Buffalo, whether by rail or canal, may be stored at the American seaboard, to be shipped at any time during the winter that may be desirable, thus placing the Montreal route at a disadvantage. The rates of marine insurance are also said to be in favour of New York. Another argument in favour of the route to New York *via* Buffalo is that the Erie Canal is open for navigation from three to four weeks later in the autumn than the St. Lawrence canals, a fact of great importance to the Western farmer who wishes to dispose of as much of his crop as possible before the close of navigation.

Montreal, Ottawa and Georgian Bay Canal.

This latest canal project is the revival of a proposal that was entertained by the Canadian Government many years ago, and upon which there was spent a good deal of money, but which was subsequently abandoned in favour of the St. Lawrence route. Mr. Macleod Stewart, ex-Mayor of Ottawa, and a man of great energy as well as of influence, is the chief promoter of the present enterprise. At his instance a company of British capitalists has been formed for the purpose of constructing and operating a system of canals to complete a through waterway from Montreal to the Great Lakes along the course of the Ottawa and Mattawa rivers, Lake Nipissing and French River to the Georgian Bay, Lake Huron —following precisely the track of the early voyageurs.

The chief advantage claimed for this route, from a commercial point of view, is that it is by far the shortest that can be devised from the Upper Lakes to the seaboard. Owing to the directness of the route it would effect a saving in distance of 450 miles over the Erie Canal route, and of 375 over the Welland and St. Lawrence route.

The total distance by the proposed route from Montreal to the waters of Lake Huron is 430 miles, requiring, it is said, the construction of only twenty-nine miles of canal, in addition to the existing canals, to complete a through waterway adapted to the navigation of vessels of 1,000 tons burthen and drawing ten feet of water. Assuming the estimated cost not to exceed $25,000,000, it is represented in the company's prospectus as an investment holding out the prospect of becoming a fairly remunerative commercial enterprise. It is further advanced in favour of the immediate prosecution of the work, that this route, being cooler and more sheltered than the lakes' route, would enable grain and cattle to be taken through in better condition; that the rate of insurance would be less; that it would render available immense natural forces in the waters of the Ottawa and its tributaries; and, especially, that owing to its distance from the international boundary it would, in case of war, be of the highest military importance, and prove of great value as a means of defence and of protection to our commerce. Provided that the necessary funds are forthcoming, there are said to be no engineering difficulties to prevent the work being completed in three years' time. On the other hand, it is alleged that a canal system limited to a draft of ten feet would not meet the present-day requirements, and could not be expected to compete successfully with one offering fourteen feet, even if the distance to be traversed would be shorter. Grain merchants, East and West, hold strongly to the opinion that the route which will bring the largest class of vessels navigating the Great Lakes to the seaboard at least expense is the route that will capture the trade. A ship canal for the Ottawa route, having twenty-five to thirty feet depth of water, with locks of 500 to 600 feet in length, would seem to offer many advantages, though in the estimation of the Deep Waterways Commission "its consideration is not now justified."

CHAPTER X.

IN THE PROVINCES OF THE DOMINION.

The History of Steam Navigation in the several Provinces of the Dominion and Newfoundland.

IN THE PROVINCE OF QUEBEC.

AMONG the names of those who were chiefly connected with the introduction and development of steam navigation in this province may be mentioned the Hon. John Molson, Messrs. John and David Torrance, and George Brush.

The founder of the Molson family and father of the steamboat enterprise in Canada came to this country from Lincolnshire, England, in 1782. Two years later he returned to Britain and raised money on his paternal estate to erect a brewery in Montreal. Subsequently he sold his English property, which enabled him to complete the Canadian enterprise that eventually grew into an extensive and lucrative business. Mr. Molson was an excellent business man and did much to advance the commercial and educational interests of his adopted country. He was President of the Bank of Montreal from June, 1826, till his death, which occurred in Montreal in 1836, in his seventy-second year. He was also an influential member of the Executive Council of Lower Canada. His son, the late Hon. John Molson, who inherited his father's enthusiasm in regard to steamboats and ship-

JOHN TORRANCE.

ping, also took a prominent part in the introduction of railways in Canada. The Molsons Bank and the William Molson Hall of McGill University are fitting memorials of the family in Montreal.

The Torrances are a "Border" family. The late

Mr. John Torrance was born at Gatehouse, in the Shire of Galloway, Scotland, June 8th, 1786. Early in the century he came to Canada, and before long established a wholesale business in Montreal and founded the eminent firm of John Torrance & Co. His elder brother Thomas had preceded him in Montreal, and was at the head of a large and lucrative business, residing at Belmont Hall, which he built, and which was at that time considered a palatial mansion. On his removal to Quebec this fine property was acquired by a member of the Molson family. Mr. David Torrance, a nephew of Mr. John Torrance, was born in New York in 1805. He came to reside in Montreal about the year 1821, and became a partner in his uncle's firm. He was a man of exceptional business capacity, energy and enterprise, and did much to advance the commercial interests of Montreal and Canada. In 1826 this firm purchased the steamboat *Hercules* and placed her on the Montreal and Quebec route, in the double capacity of a tow-boat and passenger steamer—this being the first step towards the vigorous opposition to the Molson line of steamers that ensued. They were also the first in Canada to branch out into direct trade with the East Indies and China. Mr. David Torrance died in Montreal, January 29th, 1876. His son, Mr. John Torrance, now the senior member of the firm of David Torrance & Co., was born in Montreal in August, 1835. He has had the Montreal agency of the Dominion Line of steamships for many years, and is otherwise extensively occupied in the shipping business. It may be added that after the death of Mr. John Torrance, *primus*, in 1870, the name of the firm was changed to David Torrance & Co., which it still retains.

Mr. Brush was a native of Vermont, born at Vergennes, in 1793. After some time spent in mercantile pursuits, he engaged in boat-building and navigation on Lake Champlain, and became captain of a steamer plying between St. John's and Whitehall. He afterwards had command of some of Mr. Torrance's steamers on the St. Lawrence. In 1834 he became manager of the Ottawa and Rideau Forwarding Company, and resided in Kingston until 1838, when he joined the Wards in the Eagle Foundry, Montreal, of which he became the sole proprietor in 1840. Mr. Brush died in Montreal, at the advanced age of ninety years and two months. The following extracts from memoranda left by him are interesting and valuable:

"The steam-engines for the *Swiftsure* (1813), the *Malsham* (1814), the *Car of Commerce* (1816), and the *Lady Sherbrooke* (1817), were all made by Bolton & Watt, of Soho, England, who would not allow more than *four pounds* pressure of steam; and a hand-pipe was used to feed the boilers by gravitation. The first steam-engine built in Canada was in 1819, for the *Montreal*, a small ferry-boat, of about fourteen horse-power, built by John D. Ward, at the Eagle Foundry. In 1823 the merchants of Montreal formed a stock company for the purpose of building tow-boats. I was employed by that company to build their boats.

The first (the *Hercules*) we built in Munn's shipyard, about where H. & A. Allan's office now stands. The *Hercules* was fitted with an engine of one hundred horse-power, built by J. D. Ward & Co., at the Eagle Foundry, on the Bolton & Watt low-pressure principle. Under my command the *Hercules* commenced towing vessels in May, 1824, when she towed up the

STEAMER "QUEBEC" AND CITADEL.

ship *Margaret* of Liverpool from Quebec to Montreal and up the current of St. Mary's—the first ship so towed up. Our company also built the steamers *British America*, *St. George* and *Canada*, of about 150 horse-power each."

"In 1838-39 the Imperial Government built a steam frigate here, called the *Sydenham*. It was engined by Ward, Brush & Co., with a pair of side-lever engines, and proved to be one of the fastest vessels in the Royal navy of that time."

Connected with Mr. Brush there is a good fish story, which is better authenticated than some of that class that have passed current. A pike-headed whale—the only one that is known to have visited these waters—followed some vessel up from sea into the harbour of Montreal, in September, 1823. Captain Brush rigged a boat and captured him with a harpoon. He was a beautiful specimen, measuring $39\frac{1}{2}$ feet in length, and 23 feet in circumference. His jaw-bones were for many years to be seen overarching the entrance to Gilbault's Gardens, and there are those still living who remember having seen the carcase as it lay, far too long for sensitive nostrils, on the river bank.

As already stated, Molson's *Accommodation* began to ply between Montreal and Quebec in 1809—two years later than Fulton's *Clermont* on the Hudson, and three years earlier than Bell's *Comet* on the Clyde. The *Accommodation* proved a fairly successful commercial venture, although Mr. Molson did not obtain a monopoly of the business as Mr. Fulton had done. She was soon followed by the *Swiftsure*, the *Malsham*, the *Car of Commerce*, the *John Molson*, the *Lady Sherbrooke*, and other steamboats. The last-named was 170 feet long, 34 feet beam, and 10 feet in depth, with a sixty-three horse-power side-lever engine. A much better service had now been instituted, for up to about 1818 many preferred to drive all the way from Montreal to Quebec in caleches over rough roads. Now, however, that the steamboats had comfortable cabins, and canvas awnings over their decks, they secured nearly all the through passenger traffic. About the year 1823 several powerful tow-boats were built, which also carried passengers. After these the *Waterloo* and the *John Molson* of the Molson Line, the *St. George*, the *British America* and the *Canada*, owned by John Torrance & Co., and other boats of larger dimensions, having better passenger accommodation and higher speed, followed in rapid succession. The *Waterloo* foundered in Lake St. Peter, and was replaced by the *John Bull*, a fine boat of 190 feet in length, but which was burned in 1838. The *John Bull* used too much coal to be profitable, and the saying that she made most money when lying at anchor, arose from the fact that, anchored off the city, she was repeatedly used as the official residence of the Governor-General, Lord Durham. The *Canada*, which came out in 1837, was 240 feet long, and was accounted the largest and fastest steamer then afloat in the New World. In 1840 the *Lord Sydenham* (the former *Ontario*) and the *Lady Colborne* ran as the mail boats to Quebec. About 1845 several famous boats were built—the *Rowland Hill*, Mr. Torrance's *Montreal*, Wilson Connoly's *Quebec*, the *Queen* and the *John Munn*—all upper cabin boats of high speed. The *John Munn* was longer than any previous, or, indeed, any subsequent, river steamer on the St. Lawrence, being 400 feet in length. Her boilers were placed on either guard, as the fashion then was, and a huge walking-beam in the centre. She was too large for the trade. After running a few years she was broken up, and her magnificent engines were sent to New York. The *Montreal*, also a large and fine steamer, was lost in a snow-storm near Batiscan, in November, 1853, and was replaced by the *Lord Sydenham*, afterwards lengthened to 250 feet, and renamed the *Montreal*.

The first iron steamers came into use on the St. Lawrence in 1843, namely, the *Prince Albert* and *Iron Duke*, which at that time began to ply as ferry-boats to Laprairie and St. Lambert, in connection with the Champlain and St. Lawrence railway service. These boats were designed in Scotland, sent out in segments, and were put together by Parkins, of the St. Mary Foundry, Montreal.

The Richelieu Steamboat Company, formed in 1845, commenced business by running a market boat to Sorel. In 1856 they put two small steamers on the through line to Quebec, the *Napoleon* and the *Victoria*. About this time Messrs. Tate Brothers, ship-builders, in Montreal, purchased the *Lady Colborne*, renamed her the *Crescent*, and coupling her with the *Lady Elgin*, started a fourth line of steamers to ply between Montreal and Quebec. The business had already been overdone, and this was the last straw that breaks the camel's back. The opposition had gone far enough when it had reduced the cabin fare to $1.00, including meals and stateroom, and the steerage passage to $12\frac{1}{2}$ cents! The excitement that prevailed at this time was intense. The arrival and departure of the boats at either end of the route were scenes of indescribable confusion. Vast crowds of people assembled on the wharves, while clouds of smoke issuing from the funnels and the roar of escaping steam plainly indicated that the stokers were doing their level best to burst the boilers. This vicious and ruinous opposition was brought to an end by a tragic occurrence, the burning of the steamer *Montreal*.

On a fine summer evening in June, 1857, while on her voyage from Quebec with a load of over 400 passengers, most of whom were emigrants from Scotland, who had just completed a long sea voyage, and were gazing with interest on the shores that in anticipation were to offer them happy homes, suddenly the cry of " Fire ! " was raised. Clouds of smoke burst out from between decks. A panic ensued. Groups of men and women clung to each other in despair, imploring help that was not to be found; then a wild rush, with the terrible alternative of devouring flames and the cold water below. Two hundred and fifty-three persons perished; and all the more sadly that the calamity was traced by public opinion and the press of the day to " culpable recklessness and disregard of human life." A truce to ruinous opposition ensued. An amicable arrangement was reached, by which superfluous boats were withdrawn. The bulk of the passenger business fell to the Richelieu Company, which continued for a number of years to do a lucrative trade, paying handsome annual dividends to its shareholders.

In 1875 an amalgamation was effected with the Canadian Steam Navigation Company (the old Upper Canada Line), under the name of the Richelieu and Ontario Navigation Company, which has become one of the largest enterprises of the kind in America, having a paid-up capital of $1,350,000, a fleet of twenty-four steamers, and operating a continuous line of navigation a thousand miles in length. The *Montreal* and *Quebec*, which ply between the cities from which they are named, though more than thirty years old, still have a high reputation for speed and comfort. They are each over 300 feet long, and have an average speed of about sixteen miles an hour. They have each ample sleeping accommodation for some 300 cabin passengers. They make their trips during the night. Supper on board either of these steamers is an event to be remembered.

The head office of the Richelieu and Ontario Company is in Montreal. The General Manager is Mr. C. F. Gildersleeve. Mr. Alexander Milloy, the Traffic Manager, who was born in Kintyre, Scotland, in 1822, came to Canada in 1840, when he entered the Montreal office of the Upper Canada Line of mail steamers, and continued his connection with the company, amid all its changes, until May, 1898, when he retired from the service.

On the Ottawa River.

The navigation of the Ottawa differed from that of the St. Lawrence in that its rapids were wholly impassable for boats with cargo. The necessity for canals thus became urgent. The original Grenville Canal was designed and commenced by the Royal Engineers for the Imperial Government, and was

OTTAWA RIVER STEAMER "SOVEREIGN."

completed in 1832, simultaneously with the Rideau Canal. It was enlarged by the Dominion Government a few years ago, but it is not yet of sufficient capacity to allow the free passage of the larger steamers on this route. Travellers are therefore subject to transhipment at Carillon, and are conveyed by railway to Grenville, a distance of thirteen miles, where another steamer is ready to convey them to Ottawa.

This little bit of railway is one of the oldest in Canada, and is further remarkable as being the only one of 5 feet 6 inches gauge in the country. It was purchased by the Ottawa River Navigation Company in 1859, and is operated only in connection with their steamers, not being used in winter.

The completion of the Grenville Canal in its original form opened up a new route to the West, somewhat circuitous, doubtless, but with greatly increased facilities for the transportation of merchandise, the immediate effect of which was to transfer the great bulk of west-bound traffic from the St. Lawrence route to that of the Ottawa and Rideau. About this time was formed "The Ottawa and Rideau Forwarding Company," by leading merchants in Montreal, with Mr. Cushing as manager. A few years later the forwarding business became a lucrative one, and was carried on by a number of prominent firms represented at Montreal, Prescott, Brockville and Kingston. Chief among these were the Messrs. Macpherson, Crane & Co., Hooker & Jones, Henderson & Hooker (afterwards Hooker & Holton), H. & J. Jones of Brockville, and Murray & Sanderson of Montreal. Messrs. Macpherson and Crane were easily the foremost in the enterprise, for they owned a private lock at Vaudreuil and thus held the key to the navigation of the Ottawa, and had complete control of the towage until 1841, when Captain R. W. Shepherd, then in command of the steamer *St. David*, belonging to a rival company, as the result of a clever and hazardous experiment, discovered a safe channel through the rapids at St. Ann's, which put an end to the monopoly.

Up to 1832 the long portage between Carillon and Grenville was a serious drawback to traffic, necessitating a double service of steamers and barges, one for the upper and one for the lower reach of the river. The first steamer on the upper reach seems to have been the *Union*, Captain Johnson, built in 1819, and which commenced to ply the following year between Grenville and Hull, covering the distance of sixty miles in about 24 hours! On the lower reach the *William King* began to ply about 1826 or 1827, at first commanded by Captain Johnson, afterwards by Captain De Hertel. The *St. Andrew* followed soon after. In 1828 the *Shannon*, then considered a large and powerful steamer, was built at Hawkesbury and placed on the upper route, commanded at first by Captain Grant and afterwards by Captain Kaines.

At the height of the forwarding business on the Ottawa, Macpherson & Crane owned a fleet of thirteen steamers and a large number of bateaux and barges, which were towed up the Ottawa and through the Rideau Canal to Kingston, the entire distance being 245 miles. The flotilla would make the round trip, returning *via* the St. Lawrence, in twelve or fourteen days. The steamers engaged in this service were mostly small, high-pressure boats—commonly called "puffers." At the first the noise which they made, especially the unearthly shriek of their steam-whistles, scared the natives as well as the cattle along the banks of the river. The passengers were usually accommodated in the barges in tow of the steamers, but as time went on a few of the "puffers" attained the dignity of passenger boats, and, when unencumbered with tows, made the round trip in a week. The writer well remembers making the trip in the early forties on the *Charlotte*, Captain Marshall, and a very pleasant trip it was, the chief attractions being the long chain of locks at the small village of Bytown—soon to become the beautiful capital of the Dominion; the big dam at Jones' Falls, with its retaining wall three hundred feet in thickness at the base and ninety feet high; the marvellous scenery of the Lake of the Thousand Islands, and, as the climax, what was then the novelty of shooting the rapids on a steamboat. Captain Howard informed me that the first steamer to shoot the "lost channel" of the Long Sault rapids was the old *Gildersleeve* of Mr. Hamilton's line, in command of Captain Maxwell and piloted by one Rankin. That was in 1847, and was considered a daring feat at the time, but it established the safety of the new channel which has ever since been used by the larger passenger steamers. No one, however, can form an adequate idea of the grandeur of this raging torrent who has not made the descent upon a raft; though, speaking from experience, this mode of shooting the "lost channel" is not to be recommended to persons of weak nerves.

It is said that in 1836 a steamboat named the *Thomas Mackay* plied between Quebec and Ottawa, but its journeyings seem to have been erratic and its subsequent history "lost in obscurity"—a phrase that applies in some degree, indeed, to the early history of steam on the Ottawa. The *St. David* was the only steamer that could pass through the Grenville Canal in 1841. The first truly passenger service on the Ottawa commenced in 1842 with the *Oldfield* on the lower route and the *Porcupine* on the upper. In 1846 the *Oldfield* was purchased by Captain Shepherd and others who formed a private company named the "Ottawa Steamers Company." The steamer *Ottawa Chief* was built by that company in 1848, but she was found to draw too much water, and in the following spring was chartered by Mr. Hamilton and placed on the St. Lawrence route. The *Lady Simpson*, built in 1850, was the precursor of a number of excellent steamers that have made travelling on the Ottawa popular with all classes. Among these were the *Atlas, Prince of Wales* (which ran for twenty-four years), *Queen Victoria, Dagmar, Alexandra*, etc. The reputation of the line is well sustained at present by the *Empress*, Captain Bowie, and the *Sovereign*, Captain Henry W. Shepherd, both very fine and fast steel boats of 400 and 300 tons, respectively. Other steamers in commission and employed in the local trade bear such loyal names as *Maude, Princess* and *Duchess of York*.

Captain Robert Ward Shepherd retired from active service in 1853, when he was appointed General Manager of the line. In 1864 the Steamers Company was incorporated by Act of Parliament under the name it now bears, the Ottawa River Navigation

Company, of which Mr. Shepherd was President as long as he lived. Mr. Shepherd was born at Sherringham, County Norfolk, England, in 1819. He died at his country seat at Como, Quebec, August 29th, 1895, having been for fifty-five years closely identified with

CAPTAIN R. W. SHEPHERD.

the progress of steam navigation on the Ottawa, and having earned for himself a high reputation. His brother, Captain H. W. Shepherd, who succeeded him in the command of the *Lady Simpson* in 1853, is now the commodore of the fleet—the oldest and most experienced captain on the Ottawa, who in all these years has not been chargeable for any accident to life or limb of the many thousands who have been committed to his care. The head office of the company is in Montreal, Mr. R. W. Shepherd, a son of the founder, being the Managing Director.

In the Province of Ontario.*

As already mentioned in the previous chapter, the *Frontenac* and the *Queen Charlotte* were the first two steamers in Upper Canada, launched respectively in 1816 and 1818. In 1824 another steamer was built for Hon. Robert Hamilton—the *Queenston*, of 350 tons—which was at first commanded by Captain Joseph Whitney and plied between Prescott, York and Niagara. The *Canada*, Captain Hugh Richardson, came out in 1826 and used to run from York to Niagara (36 miles) in four hours. The famous *Alciope*, of 450 tons, Captain Mackenzie, appeared in 1828, and plied with great *éclat* between Niagara, York, and Kingston, under the Hamilton flag.

The late Hon. John Hamilton, who for many years may almost be said to have controlled the passenger traffic on the Upper Canada route, commenced his connection with the steamboat business about the

* Mr. John Ross Robertson's "Landmarks of Toronto" (Toronto: 1896) contains an account of nearly all the steamboats that have plied on Lake Ontario and the Upper St. Lawrence from 1816 to 1895.

year 1830, when he built the *Great Britain*, of 700 tons, the largest vessel then on the lake. After this there was a rapid succession of steamers, and some very fine ones. The *Cobourg*, of 500 tons, Captain Macintosh, came out in 1833; the *Commodore Barrie*, 275 tons, Captain Patterson, in 1834. The *Sir Robert Peel*, 350 tons, one of the finest boats then on the lake, was seized and burned on the night of May 29th, 1838, by a gang of rebels headed by the notorious Bill Johnson. The *Queen Victoria*, Thomas Dick, commander, launched in 1837, was advertised to sail daily between Lewiston, Niagara and Toronto, connecting at Toronto with the *William IV.* for Kingston and Prescott. "This splendid fast sailing steamer is fitted up in elegant style, and is offered to the public as a speedy and safe conveyance." The *Sovereign*, 500 tons, Captain Elmsley, R.N., Captain Dick's *City of Toronto*, and the famous *Highlander*, Captain Stearns, began to run about 1840. The *Chief Justice Robinson*, Captain Wilder, the *Princess Royal*, Captain Twohey, and Captain Sutherland's *Eclipse* were all noted steamers in their day. The *Traveller* and the *William IV.*, Captain Paynter, both powerful steamers, famous also for many years, ended their careers as tow-boats, the latter being conspicuous by her four funnels.

"These steamers, and others that could be named," says one of my informants, "bring to mind good seaworthy ships, fit for any weather and commanded by able seamen. Nor was the steward's department unworthy of the vessels. As good a breakfast and dinner was served on board as could be desired." Such were some of the early steamboats in Upper Canada more than fifty years ago, for which the public are indebted to the Hon. John Hamilton, Mr. Alpheus Jones, of Prescott, Mr. Donald Bethune, of Cobourg, and Mr. Heron, of Niagara, as well as to Captains Dick, Sutherland and Richardson.

OLD "WILLIAM IV.," 1832.

Up to 1837 the lake steamers did not venture farther down than Kingston, but about that time they commenced running through the Lake of the Thousand Islands to Prescott. From that point the small steamer *Dolphin* sailed every morning for the head of the Long Sault rapids, enabling passengers to reach Montreal the same evening. The route was from Dickenson's Landing to Cornwall by stage, thence through Lake St. Francis by steamer to Coteau du Lac, thence by stage over a plank road to the Cascades, where the quaint old steamer *Chieftain* would be waiting to convey passengers to Lachine to be driven thence in a coach and six to Montreal. It was not until 1848, when the enlarged Lachine Canal was opened, that the Upper Canada steamers began to run all the rapids of the St. Lawrence as they now do.

In 1840 Mr. Hamilton had built a powerful steamer, the *Ontario*, with the expectation that she might be able to ascend the rapids, but failing in this she was sold to a Montreal firm and placed on the Quebec route. The *Ontario* descended all the rapids of the St. Lawrence safely on the 19th of October, 1840, being the first large steamer to do so. *Facile descensus!* It is not recorded that more than one steamer ever succeeded in ascending those rapids. In November, 1838, the little *Dolphin*, after four weeks of incessant toil, was towed up the Long Sault rapids with the aid of twenty yoke of oxen, besides horses, capstans and men, added to the working of her engine—the first and probably the last steamer that will ever accomplish the feat. About this time the *Iroquois*, with one large stern-wheel, was built for the purpose of stemming the swift currents between Prescott and Dickenson's Landing, but had so much difficulty in ascending the river that at Rapide Plat and other points posts were sunk at short distances along the shore to each of which she made fast in turn until she recovered her breath.

"PASSPORT," SHOWING THE RAPIDS IN HER FIFTIETH YEAR.

The completion of the canals prepared the way for a larger class of steamers between Lake Ontario and Montreal, and the " Royal Mail Line " was accordingly re-enforced. The *Passport* was built of iron on the Clyde and brought out in sections in 1847, and is still in commission and in good running order. The *Magnet*, also built of iron and on the Clyde, and in which Captain Sutherland had a large pecuniary interest, came out shortly after the *Passport*, and under the name of the *Hamilton*, in command of Captain A. J. Baker, is now, in her green old age, and with her hull as sound as a bell, performing a weekly service between Montreal and Hamilton. The *Kingston*, since named the *Algerian*, followed in 1855, and was first commanded by Captain Clarke Hamilton, now of H. M. Customs at Kingston. About this time the *Brockville*, Captain Day, the *Gildersleeve*, Captain Bowen, the *Banshee*, Captain Howard, and the *Lord Elgin*, Captain Farlinger, were well-known and favourite boats.

The fifteen years from 1840 to 1855 were the most prosperous in the history of steam navigation on Lake Ontario and the St. Lawrence. The Americans had at that time several lines of steamers plying between Ogdensburg, Oswego, Rochester and Lewiston. Some of these were large and very fine passenger steamers, such as the *United States*, the *Bay State*, the *New York*, the *Rochester*, the *Lady of the Lake*, the *Northerner*, the *Cataract*, and the *Niagara*. The Great Western Railway Company had also a fleet of splendid steamers—the *Canada*, the *America*, the

Europa and the *Western World*. At the breaking out of the American civil war, most of these vessels and some others were purchased by the United States Government and taken round to New York. Their places on the lake are now occupied by numerous screw propellers, chiefly doing a freight business, but many of them having excellent accommodation for passengers also.

The opening of the Grand Trunk Railway in 1855 proved disastrous to the steamboat interests. Mr. Hamilton, as well as many others, struggled gallantly for a time, endeavouring to stem the tide of competition with the new system of transportation, but about the year 1862 he was obliged to retire from the business which he had created and carried on successfully for thirty years. The steamers in which he had a large personal interest were sold to a joint stock company, which was named the "Canadian Steam Navigation Company." Mr. Hamilton was appointed General Manager of the new company; Sir Hugh Allan, President, and Alexander Milloy, Secretary-Treasurer. A few years later Captain Thomas Howard became Superintendent of the line, a position which he held until 1881, when he was appointed Harbour-master in Montreal. He died in Montreal on Easter Sunday, 1898. In 1875 the company united with the Richelieu Company, as already stated.

Lake Ontario.—The volume of steam traffic on Lake Ontario at the present time, though not to be compared with that on the Upper Lakes, is by no means inconsiderable. From the official "Report of

Trade and Navigation of the Dominion for 1895," the arrival and departure of steamers at eighteen ports of entry on Lake Ontario, either as coasting vessels or as trading with the United States, was 17,558, and an aggregate of 6,443,443 registered tonnage; to which must be added the large amount of steam shipping that frequents the harbours on the American side of the lake, as at Lewiston, Oswego, Sackett's Harbour, Cape Vincent, and that descends the St. Lawrence to Ogdensburg. Niagara heads the list on the Canadian side with 3,198 arrivals and departures, and 1,581,643 tonnage. Toronto, with 3,844 arrivals and departures, counts for 1,569,123 steam tonnage; Kingston stands third, with 3,563 vessels, and 882,414 tonnage. Hamilton is represented by 427,100 tonnage. After these come Belleville, Picton, Cobourg, Port Hope, Descronto and Port Dalhousie, in the order named, and eight other smaller ports, each contributing its quota.

Toronto is largely interested in steam navigation. Not to speak of numerous steam yachts, ferry steamers and tug-boats, it controls a large passenger traffic. The Niagara Navigation Company of Toronto has three very fine steamers running to Niagara and Lewiston—the *Chicora*, *Chippewa* and *Corona*. The *Chicora* was built in England, as a "blockade runner," more than thirty years ago, but the civil war was ended before she reached this side of the Atlantic. She is an iron side-wheel vessel of 518 tons, with a rakish, Old-Country look about her. The *Chippewa*, built at Hamilton, Ont., in 1893, is a very fine paddle-wheel steamer of 850 tons, modelled somewhat after the Hudson River boats, with a conspicuous walking-beam. The latest addition to the fleet is the *Corona*, launched in May, 1896, from the noted ship-building yard of the Polsons, Toronto, which takes the place of the *Cibola*, a Clyde-built steel steamer, put together by the Rathbun Company, Deseronto, in 1887, and which was burned at Lewiston in 1895. The *Corona* is claimed by her owners to be "a model of marine architecture, and one of the finest day-steamers in the world!" Though only 277 feet long, and 32 feet beam (59 feet over the guards), she carries nearly two thousand passengers. The hull is constructed of open hearth steel. The engine is of the inclined compound condensing type, and develops nearly two thousand indicated horse-power. The mechanical fittings are all of the most approved kind, and the internal arrangements highly artistic.

The Hamilton Steamboat Company has two fine powerful screw steamers, the *Macassa* and *Modjeska*, plying between Hamilton and Toronto. Both were built on the Clyde, and have been very successful financially, and also as seaworthy, fast sailing vessels. Kingston, which occupies an important position at the foot of the lake and head of the river navigation, owns a fleet of no less than forty-six steamers, and is the headquarters of half a dozen steamboat companies, some of which are largely interested in the Lake Superior trade, while others connect Kingston with ports on the Bay of Quinte, Rochester and Cape Vincent, N.Y., and Gananoque and the Thousand

Islands. The *James Swift* plies between Kingston and Ottawa, *via* the Rideau Canal. The *Passport*, the oldest steamer now afloat in Canada, is registered at Kingston, and was built, as already stated, in 1847.

HON. JOHN HAMILTON.

The Hon. John Hamilton, whose name is so intimately associated with the rise and progress of steam navigation in Western Canada, was born at Queenston, Ontario, in 1802—the seventh and youngest son of the Hon. Robert Hamilton, formerly of Edinburgh.

One of the sons founded the city of Hamilton, another attained distinction in the medical profession. John devoted the greater part of his life to the development of commerce between Montreal and the cities and towns bordering on Lake Ontario, having his headquarters at Kingston. Mr. Hamilton was a man of fine presence and highly accomplished; was called to the Legislative Council of Upper Canada by Sir John Colborne in 1831, and to the Senate of the new Dominion, by writ of Her Majesty's sign-manual, in 1867. He was an influential member of the Presbyterian Church, and many years chairman of the Board of Trustees of Queen's College, Kingston. He died in 1882.

IN MANITOBA.*

The first steamer to ply on the Red River was brought in pieces across the country from a tributary of the Mississippi, and rebuilt at Georgetown, a small place some twenty miles north of the present town of Moorhead. The boat was called, before its transportation, the *Anson Northrup*, and was afterwards known as the *Pioneer*. She began her career on the Red River in 1859, and in that year took a cargo to Fort Garry. She was the joint property of the Hudson's Bay Company and Messrs. J. C. and H. C. Burbank & Co., of St. Paul, Minnesota. (A cut of this steamer may be seen in a book called "The Winnipeg Country," published by Cupples, Upham & Co., Boston.)

* From notes by Rev. Professor Bryce, LL.D., of Winnipeg.

The next steamer was the *International*, built at Georgetown, in 1861, for the Hudson's Bay Company, at a cost of about $20,000. Her length was 160 feet, breadth 30 feet, depth (from the water-line to the ceiling of her upper saloon) 20 feet, and her registered tonnage was $133\frac{1}{3}$ tons. She was found to be too large for the Red River navigation. The same company's steamer, the *Northcote*, commenced to ply on the Saskatchewan about 1875. In 1878 there were running on the waters of Manitoba seventeen steamers, among which were the *Manitoba, Dakota, Selkirk, Swallow, Minnesota, Prince Rupert, Keewatin*, etc.

The Hudson's Bay Company at that time owned a propeller which ran on Lake Winnipeg to the portage at the mouth of the Saskatchewan, where connection was made with the *Northcote* and a steel-built steamer, the *Lilly*. This company had also another steamer plying on the Red River, named the *Chief Commissioner*.

Since the opening of the country by railways the navigation of the Upper Red River and the Assiniboine has been of small account, but below Selkirk there is still a considerable trade carried on. There are at least half a dozen companies interested in the navigation of these waters. The North-West Navigation Company runs three steamers, the *Princess*, 350 tons; the *Red River*, 200 tons; the *Marquette*, 160 tons, and a number of barges. The Selkirk Fish Company owns the *Sultana*, of 200 tons; the Manitoba Fish Company has the *City of Selkirk*, of 160 tons. Besides these there is a numerous fleet of steam-tugs and barges. In all there are some fifty steamers on these inland waters. During the palmy days of Red River transportation the leading name was that of Norman W. Kittson, at that time of St. Paul, Minnesota, but formerly a trader of the old Red River settlement, who was often familiarly called "Commodore Kittson."

In British Columbia.*

The pioneer steamship of the Northern Pacific was the *Beaver*, whose history from first to last was a very romantic one. This vessel was built at Blackwall, on the Thames, by Messrs. Green, Wigram and Green, for the Hudson's Bay Company, and was launched in 1835 in the presence of 150,000 spectators, including William IV. and many of the English nobility. Cheers from thousands again greeted her in answer to the farewell salute of her guns when she sailed away for the New World. The *Beaver* was a side-wheel steamer, 101.4 feet long, 20 feet beam, and 11 feet deep; tonnage, 109. Her machinery, made by Boulton & Watt, was placed in position, but the paddle-wheels were not attached. She was rigged as a brig, and on August 27th sailed for the Pacific under canvas, in command of Captain Home, with the barque *Columbia* as her consort. On March 19th, 1836, the *Beaver* dropped anchor at the

* Mr. J. A. Thomson, Inspector of Steamboats for British Columbia, furnished the information contained in these notes.

mouth of the Columbia River, having made the voyage in 204 days. In her log-book it is recorded on May 16th: "Carpenters stripping paddle-wheels. At 4 p.m. engineers got up steam, tried the engines,

THE LAST OF THE OLD "BEAVER."

and found to answer very well; at 5 o'clock, came to anchor, and moored in our old berth; at 8 o'clock all hands were mustered to 'splice the main brace'"—a nautical phrase used in reference to the custom, less common now than then, of celebrating particular events by serving out a liberal supply of rum. The *Beaver* went into service without delay, running up and down the coast, in and out of every bay, river and inlet between Puget Sound and Alaska, collecting furs and carrying goods for the company's posts.

On March 13th, 1843, the *Beaver* arrived at Camosun with Factor Douglas and some of the Hudson's Bay Company's people to found the Fort Victoria, and the first salute which echoed in what is now Victoria harbour, was fired on the 13th of June, when the fort was finished and the company's flag hoisted.* "The old steamer *Beaver*," as she was called, continued her rounds under different owners with remarkable regularity and success until the fatal trip in July, 1888, when she went on the rocks near the entrance to Vancouver harbour, and was totally wrecked.

It was fourteen years after the arrival of the *Beaver* before much effort was made at steamboating in these waters. About that time several small steamers were built on the Columbia River. In 1852 the Hudson's Bay Company had another vessel built at Blackwall: this was the *Otter*, a screw steamer of

---

* Vancouver Island was at that time a British possession—leased to the Hudson's Bay Company. When the lease expired, in 1859, the Island was made a Crown colony, and the old fort, with the large cattle farm attached to it, became the site of the beautiful city of Victoria, with its fine streets, electric railways, magnificent public buildings, palatial residences, a population of 23,000, and real estate valued at $20,000,000. The Island and British Columbia were made one Province in 1866, and entered the Dominion in 1871.

220 tons, with a pair of condensing engines by Penn, of Greenwich, which took the first prize at the London Exhibition in 1851. The *Otter* left London in January 1853, and arrived at Victoria five months later. The year 1858 witnessed a boom in steam navigation, consequent upon the rush and wild excitement of gold-seekers to the Fraser River and Cariboo. "The *Surprise* first woke the echoes in the grand mountain gorges in the wild regions of Fort Hope with the

THE STERNWHEELER "NELSON," AT NELSON, B.C.

shrill scream of the steam-whistle, and astonished the natives with her wondrous power in breasting successfully the fierce current of the now world-renowned Fraser. That wild and unearthly yell of the imprisoned steam escaping into the free air of heaven must have astonished the denizens of those mountain fastnesses and startled man and beast into the belief that some uncanny visitor, not of earth, had dropped in upon their solitude." The *Surprise* was followed by a fleet of small steamboats built in the United States.

Among those were the *Ranger* and *Maria*—mere steam launches of about 40 feet in length. The *Maria* was brought up from San Francisco in a barge. The first boat built in British Columbia was the *Governor Douglas*, a good-sized stern-wheeler which commenced to ply between Victoria and the Fraser River in 1859. Among the other notable boats were the *Seabird* and the *Eliza Anderson*. The former carried immense crowds, but drew too much water for the river trade. The latter was a side-wheeler, built in Portland, 140 feet long, and of registered tonnage, 279. On her arrival at Victoria in 1859 she commenced a career of money-making which has seldom been equalled. After these appeared the *Umatilla, Enterprise* and *Colonel Moody*, the last-named being the fastest yet built for this route. All the light-draught boats were then, as they are now, stern-wheelers. About this time another and larger vessel arrived from London, the *Labouchere*, a side-wheel steamer, of 680 tons register, 202 feet long, 28 feet beam, and 15 feet hold. She continued running up north until 1865, when she was granted a subsidy of $1,500 a trip to carry mails between Victoria and San Francisco, but was lost on her first voyage. In 1861 more steamboats were built than in any previous year. Nearly a dozen were added to those already plying on the rivers and lakes, and the subsequent progress in steam navigation was continuous. The entrance of mining prospectors into the Kootenay country in 1886 led to the necessity of increased transportation on the Columbia River, which has gone on increasing until now on that river and the Kootenay lakes there are some of the finest river steamers in the Dominion, fitted with every comfort and appliance that experience can suggest. The development of the coast wise trade has also led to the building of special steamers both in British Columbia and also in England. The coal mines at Nanaimo and the Comox district also find employment for a large quantity of steam tonnage.* The aggregate amount at the four ports of Victoria, Vancouver, Nanaimo, and Westminster for 1895 was: Arrivals, 1,496,409 tons; departures, 1,513,233 tons. There are at present registered in British Columbia 161 steamboats with a tonnage of 24,153.

Besides the inland steamers there are coasting lines from Victoria and Vancouver to Portland and San Francisco, and to Puget Sound and Alaska. There are also four regular lines of steamships to Japan and China, namely, the Canadian Pacific Steamship Company, with its beautiful fleet of "Empress" steamers; the Northern Pacific Steamship Company; the Oregon R. R. and Navigation Company, and the Nipon Yunen Kaisha of Japan. There is also the direct line of steamers to Australia elsewhere referred to. The number of vessels in the different lines is uncertain, as they are increased by chartered boats whenever there is much freight moving.

* Since these lines were penned the rush to the Klondike has given an immense impetus to the steamboat business of British Columbia.

In Nova Scotia.*

The harbour of Halifax is one of the finest in the world. It is easy of access and open all the year round. It is nearly six hundred miles nearer to Liverpool than is New York, and has therefore many advantages to offer as a point of arrival and departure for ocean steamers. It is the centre of an extensive local and coasting trade, in which a large number of both steamers and sailing vessels are employed. The number of arrivals of sea-going vessels in 1895 was 978, with a gross tonnage of 627,572 tons; the number of arrivals of coasting vessels was 3,651, of which 496 were steamers, with a tonnage of 153,790 tons. The number of steamers registered in the port is 55, with a gross tonnage of 10,912 tons. The steam tonnage which entered the port in 1896 was 212,085; the clearances were 229,653 tons.

The first steamer to enter this renowned harbour was the *Royal William* (Captain John Jones, R.N.), from Quebec, August 24th, 1831, which arrived here on the morning of the 31st and was welcomed with great *éclat*. The trip was made in six days and a half, including two days' detention at Miramichi. The cabin fare was £6 5s., including meals and berths. Having been built for this trade, the *Royal William* made a number of successful voyages between Quebec and Halifax, calling at intermediate ports previous to her historic voyage across the Atlantic, which was to proclaim her the pioneer of ocean steam navigation!

* From notes by Rev. Robert Murray, Halifax.

The Cunard Line commenced to call at Halifax fortnightly *en route* to Boston, in 1840. The *Britannia* was the first of that famous fleet to enter the harbour of Halifax. This arrangement did not last very long, however, for, on making New York their western terminus, the Cunarders gave "the finest harbour" the go-by, never to return except in cases of emergency. There are, however, some fifteen or sixteen lines of steamers plying regularly from Halifax to Britain, the United States, the West Indies, South America, Newfoundland, and Canadian ports. During the winter months the Beaver Line, carrying the Canadian mails, calls here weekly *en route* from St. John, N.B., to Liverpool. The Allan Line from Liverpool to Philadelphia, *via* Newfoundland, touches here once a fortnight going and coming. The Furness Line has excellent steamers sailing fortnightly from London to Newfoundland and Halifax. The Canada and Newfoundland Line also maintains a good service from Halifax to St. John's, Liverpool and London; the Jones Line to Jamaica; the Pickford and Black Line to Bermuda and the West Indies; the Musgrave Line to Havana. The Red Cross Line from New York to Newfoundland calls here; besides, a number of coasting steamers to Cape Breton, Newfoundland, Yarmouth, Bridgewater, St. Pierre and other places call at Halifax, while the Canada Atlantic and Plant Line supplies a direct route to Boston and all points in the United States.

Many "tramp" steamers call at Halifax with freight or for freight. Many call for coal. Many a storm-tossed mariner is glad to make for Halifax and to find in it a secure harbour of refuge, with all needful appliances for refitting a battered ship. The whole coast of Nova Scotia, indeed, is indented with harbours of refuge, which are the resorts of large numbers of sailing craft. The graving-dock at Halifax is the largest on this continent. It was completed in 1889 by a private company, subsidized by the Imperial and Federal Governments and the city of Halifax to the extent of about $30,000. It is 585 feet in length, $89\frac{1}{4}$ feet wide at the entrance, and has 30 feet of water on the sills. It is adapted for steamships of the *Teutonic* class, but is 35 feet too short for the *Lucania*. A few months ago it had the honour of accommodating within its walls the *Indiana*, one of the largest of the United States ships of war, sent here for repairs. There are three other graving-docks, the property of the Dominion Government, as follows :*

At Esquimalt, B.C., built in 1886, 430 x 65 x $26\frac{1}{2}$ feet.
" Kingston, Ont.,   "   1871, 280 x 55 x $16\frac{1}{2}$ "
" Levis, Que.,   "   1887, 445 x 62 x $26\frac{1}{2}$ "

---

* The largest graving-dock in the world is said to be the one built for the Clyde Trust at Govan, on the Clyde, and recently opened. It is 880 feet long, 115 feet wide and has $26\frac{1}{2}$ feet of water on the sill. The Clyde Trust are evidently looking ahead. There may be no ships of 850 feet in sight at the moment, but there is no telling how soon there will be. The Govan dock is ready for them. In the meantime it has been partitioned off into two parts by still gates, the outer division being 460 feet in length, and the inner, 420 feet.

In New Brunswick.*

The first steamboat in New Brunswick, the *General Smyth*, was launched from the yard of John Lawton, Portland, St. John, in April, 1816. Her owners were John Ward, Hugh Johnson, sen., Lauchlan Donaldson, J. C. F. Bremner, of St. John, and Robert Smith, of Fredericton. This vessel was run between St. John and Fredericton, making the round trip in a week. She started from St. John on her first trip, May 13th, 1816. She was a paddle boat. No official description of her is extant, as the registry book of that date was burned in the great fire of 1877. Later steamboats on this route were the *St. George, John Ward, Fredericton, St. John, Forest Queen, Heather Bell, Olive, Prince Arthur, David Weston, Rothsay* (which afterwards ran between Montreal and Quebec), the *Fawn* and *May Queen*.

The second steamer, the *St. George*, was launched on April 23rd, 1895, from the yard of John Owens, at Portland, St. John. Her owners were John and Charles Ward, of St. John; Jedediah Slason and James Segee, of Fredericton—the last-named being the first master of the vessel. Her tonnage was $204\frac{17}{}$; length, 105 feet; greatest breadth, 24 feet $6\frac{1}{2}$ inches; depth of hold, 8 feet 6 inches. She had one mast, a standing bowsprit, square stern, and was carvel built. She had a copper boiler, and, like the *General Smyth*, made one trip each way between

* Information furnished by Mr. Keith A. Barber, of H. M. Customs, St. John, N.B.

Fredericton and St. John in a week. The *Victoria*, the first steam ferry-boat between St. John and Carleton, commenced running September 5th, 1839.

The pioneer steamboat on the Bay of Fundy was the *St. John*, built at Deer Island, N.B., in 1826. In her was placed the machinery of the *General Smyth*. Her tonnage was $87\frac{84}{94}$; length, 89 feet; breadth, 18 feet; depth, 8 feet. Later boats on this route were the *Royal Tar, Fairy Queen, Maid of Erin, Pilot, Emperor, Commodore, Empress, Scud, Secret* and *City of Monticello*. The steamers at present running from St. John are: to Digby, the steel paddle SS. *Prince Rupert*, 620 tons, having a speed of $18\frac{7}{8}$ knots; to Windsor and Hantsport, N.S., the *Hiawatha*, 148 tons; to Yarmouth, N.S, the *Alpha*, 211 tons; to Grand Manan, the *Flushing*, 174 tons.

The first New Brunswick steamer to ply between St. John and Boston was the *Royal Tar*, $256\frac{89}{94}$ tons, Thomas Reed, master, built at Carleton in 1835. She was burned in Penobscot Bay, October 25th, 1836, on her voyage to Portland, Maine, when thirty-two lives were lost; also a whole menagerie with elephants, horses, etc. This service is now performed daily by the International Steamship Company of Portland, Maine, who have three splendid steamers on the route —the *State of Maine*, 818 tons; the *Cumberland*, 896 tons, and the *St. Croix*, 1,064 tons. On the River St. John there are eight passenger steamers and eleven tug-boats. A large number of tugs also ply on the harbour. The number of steamers that entered the port during the year ending June 30th, 1897, was

823, aggregating 609,319 tons. Of these, 359 were ocean and 464 coasting steamers. The lines of ocean steamers plying to and from St. John during the winter of 1897-98 were : the Furness Line, to London and to the West Indies; the Beaver Line, carrying Her Majesty's mails to Liverpool, *via* Halifax and Moville; the Allan Line and William Thomson & Co.'s boats to London; the Donaldson Line, to Glasgow, and the Head Line, to Belfast and Dublin.

Many advantages are claimed for St. John as a winter port for the Dominion. In point of distance from Liverpool it has the advantage over Portland of 80 miles, and over New York of 450 miles. Halifax is nearer England by 200 miles, but the land carriage from the West is much greater. St. John is the centre of an extensive lumber business. It is connected with Western Canada by both the Intercolonial and Canadian Pacific railways. The approach to the harbour is said to be free from fogs in the winter months, and ice is altogether unknown in the Bay of Fundy. Large sums of money have been expended during the last few years in improving the export facilities, and the lieges of St John see no reason why this port should not become the Canadian winter terminus of the coming "Fast Line."

Captain W. L. Waring, the Inspector of Steamboats in New Brunswick and Prince Edward Island, claims that the invention and application of the compound steam-engine, which has done so much towards the increase of power and lessening the amount of fuel for its production, belongs rightfully to Canada. Though experiments had been made in using steam twice for the same engine, it was only in 1856 that John Elder, of the Fairfield Ship-building Company on the Clyde, reduced it to a practical success in Britain, and it was not until 1870 that it came into general use. Captain Waring states that the steamer *Reindeer*, 129 feet 9 inches long, 13 feet 8 inches wide and 8 feet deep, was built by Thomas Prichard at Fredericton, N.B., and launched April 20th, 1845, and that she was fitted with compound engines, the diameter of the high-pressure cylinder being 17 inches, of the low-pressure cylinder 32 inches, and the length of stroke 4 feet 9 inches. "This," says Captain Waring, "was the pioneer steamboat with engines using steam the second time. For the first four or five years she was not a success. While the principle was good, the machinery was defective, and between the incredulity of the people and the defects in the machinery she was near being laid up as a failure. After a thorough overhaul, it was demonstrated on her trial trip—the writer being on board—that she was a success, in proof of which the owners of the steamers on the St. John River bought her at an advance of four times what they offered for her in the fall." It is added that the *Reindeer's* machinery was placed in a new boat called the *Antelope*, which proved a great success, being very fast. It was next placed in the *Admiral*, where it now is, the original compound engine of 1845.

Honour to whom Honour! Mr. Barber states that

the first steam fog-whistle in the world was started on Partridge Island, at the entrance of St. John harbour, in 1860, under the superintendency of Mr. T. T. Vernon Smith. " The whistle was made by Mr. James Fleming, of St. John, in 1859."

In Prince Edward Island.*

The smallest of the provinces of the Dominion and the last to enter Confederation, in 1873, has long been noted for its marine enterprise, its ship-building, and its fisheries. As many as a hundred sea-going vessels have been built there in a single year; but iron and steel in these days have so largely superseded wood, this branch of industry has greatly decreased in Prince Edward Island, which modestly claims not much more than 2 per cent. of the registered steam tonnage of the Dominion of Canada.

The first steamer to enter any port in Prince Edward Island was a tug-boat, built in Pictou for the Albion Mines Coal Company, and named after the then manager, *Richard Smith*. She brought over a party of excursionists to Charlottetown, on August 5th, 1830, and returned the same day. On September 7th, 1831, the famous *Royal William*, on her first return voyage from Halifax to Quebec, called at Charlottetown, but as the merchants of that place declined to purchase the fifty shares of stock in the

* Information supplied by Mr. W. F. Hales, of Charlottetown.

new enterprise, which they had been offered conditionally, she called there no more. On May 11th, 1832, a steamer named the *Pocahontas*, built in Pictou, commenced to ply between that port and Charlottetown, about fifty miles distant, under arrangement with the post-office authorities. This vessel was followed at successive intervals by the *Cape Breton*, the *St. George*, the *Rose*, and the *Rosebud*, the last three being owned on the Island. A fine steamer, the *Lady Marchant*, owned in Richibucto, also made Charlottetown a port of call. There were many periods, however, between these steamers when communication with the Island had to be kept up by sailing schooners, until about 1852, when a regular service was commenced by the *Fairy Queen* and the *Westmoreland*, between Point du Chene and Summerside, and thence to Charlottetown and Pictou.

In 1863 the Prince Edward Island Steam Navigation Company was organized, and the steamer *Heather Belle*, built in Charlottetown, began the service in 1864, followed by the *Princess of Wales*, built at St. John, N.B. The *St. Lawrence* was added in 1868. With these three steamers a regular service was maintained between Miramichi, Richibucto, Point du Chene, Summerside, Charlottetown, Brulé and Pictou, until the railway was opened to Pictou, when the service was extended to Port Hood and Hawkesbury, on the Gut of Canso, and to Georgetown and Murray Harbour on the Island. Again, on the completion of the Cape Breton railway and the extension of the Island railway to Georgetown, the service was changed to a daily route between Charlottetown and Pictou, and Summerside and Point du Chene, as at present. The new steamers, *Northumberland* and *Princess*, are scarcely surpassed for the work they have to do by any steamers in Canada, and the company are able to show a record which is probably unique—that during thirty-three years not an accident has occurred by which a person or a package of freight has been injured.

Some years ago the North Atlantic Steamship Company was organized at Charlottetown, with a view of establishing a direct trade with the Old Country. The fleet consisted of one steamer only, the *Prince Edward*, and as the enterprise did not prove self-sustaining, after having run for several seasons the vessel was sold at a considerable loss to the shareholders.

THE WINTER FERRY.

Prince Edward Island, lying in the southern part of the Gulf of St. Lawrence, is separated from the mainland by the Strait of Northumberland, which at its narrowest point is about nine miles wide. Owing to the accumulation of ice by which this strait is obstructed in winter, communication with the Island at that season of the year has always been attended with difficulty and not unfrequently with danger. For many years the only conveyance for mails and passengers in winter was by means of open boats or canoes manned by expert boatmen. Latterly these boats, most of which now belong to the Government of Canada, have been greatly improved. They now make the passage never less than three together, each manned by five able men, and the fleet under the charge of an experienced ice-captain. If large ice-fields should be jammed between capes Tormentine and Traverse, the crossing may be made without putting the boats into the water at all—the men, assisted by the male passengers, hauling the boats over the ice by straps fastened to the gunwales. When the ice is good the passage may be made in three or four hours. At other times lanes of open water occur into which the boats are launched and rowed as far as practicable. If there is much "lolly" to work through, this entails great loss of time and labour. Or the ice may be very rough and hummocky, which makes the crossing difficult and tedious. When overtaken by a snow-storm there is danger of losing the bearings and of travelling in the wrong direction. There have been occasions when parties have been out all night and nearly perished; but since the Government has taken charge of the ferry better regulations are in force. Each boat carries a fixed number of passengers and a limited amount of mail and baggage. This, with carrying compasses, provisions, and proper fur wraps, has greatly improved the service.

The ice attached to the shores on either side of the strait extends about one mile, leaving seven miles for the ferry, but owing to the run of the tide—about four miles an hour—which carries with it, to and fro, huge masses of ice, often closely packed, the actual distance traversed by the boat is greatly increased. Horses and sleighs await the arrival of the boats at the board-ice on either side, when the passengers and mails are conveyed to the boat-sheds. For about two months every winter this boat service proves the quickest and most reliable means of crossing, and it is likely to remain so.

At the time of Confederation the Dominion Government guaranteed to provide the Island with a steam ferry service. The first effort to carry out the agreement was made by employing an old steamer, the *Albert*, to run between Pictou and Georgetown, but she had not sufficient power to force her way through the ice. In the meantime the *Northern Light* was being built at Quebec—a vessel of considerable power and extraordinary shape. She drew nineteen feet aft, and it was intended that her keel, forward, should be above the water-line, but owing to a miscalculation as to her displacement, it proved to be some two feet below, and this spoiled her for ice-breaking; but on the whole she did good service from 1876 to 1888, although she was often "frozen in," and was for several weeks at a time fast in the ice when full of passengers.

The *Stanley*, which succeeded the *Northern Light*, was built in 1888 at Govan on the Clyde, after the model of similar ice-steamers in Norway and Sweden. She has done excellent service, and her powers of breaking ice and separating large floes must be seen to be understood or believed. That she has not been able to keep up continuous communication does not surprise those who know what the Gulf is at some seasons of the year. She has made passages when it seemed futile to expect it; and while she has been imprisoned in the ice for as much as three weeks at a time, she has made the voyage from Pictou to Georgetown—40 miles—in two hours and a half.

"STANLEY," WINTER FERRY-BOAT TO PRINCE EDWARD ISLAND, 1881.

During the season 1894-95 the *Stanley* carried 1,600 passengers. Her earnings were $9,266.92; the cost of her repairs and maintenance was $28,179.32.

The *Stanley* is built throughout of Siemens-Martin steel. Her dimensions are: length, 207 feet; breadth, 32 feet; depth, 20 feet 3 inches. She is a screw boat of 914 tons gross, and 300 horse-power, and attains a speed of nearly 15 knots in clear water. She is so constructed that she runs up on heavy ice, breaking it with her sheer weight. At times she has passed through what is called "shoved ice," eight feet in thickness. She has good state room accommodation for about fifty cabin passengers, and is in every way a very efficient, powerful and staunch boat.

In the spring and fall of the year the *Stanley* is employed in the Coast Buoy service; in summer she takes her place in the Fisheries' Protection fleet, and proves herself a smart and formidable cruiser and a terror to evil-doers. She commences the winter mail service from Charlottetown to Pictou about the first of December, and about Christmas, when the Charlottetown harbour is frozen over, she takes up the route from Pictou to Georgetown, at the eastern end of Prince Edward Island. When she is imprisoned in the ice, as frequently happens, the mails and passengers are taken by the open boats in manner above described. From February 8th to April 12th, 1895, when the *Stanley* was laid up for repairs, the ice-boat service carried 3,497 mail bags, 458 pounds of baggage, 76 pounds of express goods, 9 passengers, and 77 "strap-passengers."

Dominion Steamers.

In connection with the Lighthouse and Buoy service and the Fisheries' Protection the Canadian Government employs fourteen steamers and three sailing vessels. The aggregate gross tonnage of the steamers is 5,589 tons. Of these the *Stanley* is the largest, after which come the *Newfield*, 785 tons; the *Aberdeen*, 674 tons; the *Acadia*, 526 tons—all of Halifax; the *Lansdowne*, 680 tons, of St. John, N.B.; the *Quadra*, 573 tons, of Victoria, B.C.; *La Canadienne*, 372 tons, of Quebec, etc., etc.

NEWFOUNDLAND.[*]

The history of steam navigation in this province begins with the year 1840, when Her Majesty's ship *Spitfire*—a paddle steamer—entered the harbour of St. John's with a detachment of soldiers to strengthen the garrison. In 1842 the steamship *John McAdam* visited St. John's, and a number of ladies and gentlemen made excursions in her to Conception and Trinity bays, startling the natives by the sight of a vessel walking the waters without the aid of sails or oars. In 1844 the Government arranged with the owners of the steamship *North American* to carry mails and passengers regularly between St. John's and Halifax. When this vessel first entered the harbour, with her huge walking-beam and a figurehead of an Indian, painted white, half of the population of the city crowded the wharves to see her. She had made the run from Halifax in sixty hours. Soon after this a contract was made with the Cunard Company for a mail service between St. John's and Halifax, fortnightly in summer and monthly during the winter

[*] By the kindness of Rev. Moses Harvey, D.D., of St. John's.

months. In 1873 direct steam communication with England and America was established by arrangement with the Allan Line for the conveyance of mails, passengers and goods, fortnightly during nine months of the year and monthly during the remaining months, though at a later date fortnightly trips were made all the year round.

At the present time there are five regular lines of steamships sailing from St. John's—the Allan Line, the Canadian and Newfoundland Steamship Company, the Red Cross Line, the Black Diamond and the Ross Lines. Besides these, a steamer plies regularly between Halifax and the western ports of Newfoundland; and two local steamers ply between St. John's and the principal ports north, south and west.

The total number of steamers registered in St. John's is thirty-two, with a gross tonnage of 9,272 tons. About 1,500 vessels arrive and depart annually from the several ports of Newfoundland. The sealing fleet comprises some twenty steamers, with a united tonnage of 6,230 tons, and crews numbering 4,680 men. The first steamers to engage in the seal fishing were the *Bloodhound* and the *Wolf* in 1862. The former arrived with 3,000 seals, and the latter with only 1,300. The largest catch of seals recorded was in 1844, when 685,530 were captured. The codfishing industry is carried on by sailing schooners. The annual catch in the Newfoundland waters is about 1,350,000 quintals of 112 pounds. But the total amount of cod caught in North American waters is estimated at 3,700,000 quintals annually. Allowing fifty fish to a quintal, we have the enormous number of 185,000,000 fish caught every year. And still they continue to multiply and replenish the sea!

As yet no steamers have been built in Newfoundland.

General Summary.

The total number of vessels on the registry books of the Dominion on December 31st, 1896, was 7,279, with a gross tonnage of 789,299 tons. Of that number 1,762 were steamboats, with a gross tonnage of 251,176 tons.* The steam tonnage of the Dominion is divided about as follows: Ontario, 41.1 per cent.; Quebec, 32.3 per cent; British Columbia, 10 per cent.; Nova Scotia, 7.9 per cent.; New Brunswick, 3.8 per cent.; Manitoba, 2.6 per cent.; Prince Edward Island 2 per cent.

The total number of steamers registered and enrolled in the United States in 1896 (including steam yachts, barges, etc), was 6,595 vessels, with a tonnage of 2,307,208 gross tons.†

The total number of steam vessels in the United Kingdom of Great Britain and Ireland, over 100 tons gross, recorded in Lloyds Register for 1896-97, was 6,508; their gross tonnage was 9,968,573 and their net tonnage, 6,143,282. Including the British Colonies, the number of steam vessels is 7,373 and their

* "Statistical Year Book of Canada, 1896," p. 280.

† "Report U. S. Commissioner on Navigation, 1896," p. 201.

gross tonnage, 10,508,443 tons.* Of these only about 420 are built of wood, 3,883 are built of iron and the rest of steel.

The World's Steamers.

According to Lloyds Register above quoted, the total number of steam vessels, over 100 tons, in the world in 1897 was 13,652, and their gross tonnage, 17,737,825 tons. The number of wooden steamers was 1,163; of iron, 7,099, and 5,390 of steel.

The British Empire owns 54 per cent. of the entire merchant marine tonnage of the world, estimated by Lloyds at 25,614,089 tons gross; she owns 62 per cent. of the entire merchant marine steam tonnage.

If to these figures were added the number of steam vessels in the navies of the world, the grand total would be very largely increased. The British navy alone would increase the number of vessels by 700 at least, and the tonnage by more than 1,500,000 tons.

Conclusion.

Reliable statistics are not easily found and are often accounted dry reading. From a variety of causes, figures are peculiarly prone to err. But whatever may be thought of the merely numerical argument which has almost unavoidably been introduced in these pages, the indisputable fact remains, that of all the triumphs of mind over matter in this nineteenth

---

* "Report U. S. Commissioner on Navigation, 1896," p. 127.

century nothing has contributed more to the advancement of civilization and the spread of Christianity, to the wealth of nations and the convenience and comfort of the human race, than the marvellous development of steam navigation which will ever be identified with the history of the illustrious reign of Her Majesty QUEEN VICTORIA.

APPENDICES.

I.

CAPTAIN JOHN ERICSSON.

The name and fame of the inventor of the screw propeller are less widely known in Britain than in America, and in neither country, perhaps, has full justice been done to his memory. As a mechanical genius, he was one of the most remarkable men of his time, and did much to promote the development of steam navigation.

Ericsson was born in the Province of Vermeland, in Sweden, in the year 1803. Coming to England in 1826, he entered into partnership with Braithwaite, a noted mechanician, in London, and there and then entered upon his remarkable career as an inventor. In 1836 he married Amelia, daughter of Mr. John Byam, second son of Sir John Byam. Accompanied by his wife, he came to the United States, arriving at New York, in the *British Queen*, November 2nd, 1839. His wife, however, soon afterwards returned to England, and during the rest of their lives, "by an amicable arrangement," the Atlantic rolled between.

Before leaving England, Ericsson had already patented a number of his inventions. One of the first of these was a machine for compressing air, a discovery which has since proved valuable in the construction of long tunnels and in many other ways. The introduction of his system of artificial draught was the key-note of the principle on which rapid locomotion chiefly depends. He electrified London with his steam fire-engine, but the conservative authorities would not countenance "a machine that consumed so much water!" In 1829 he entered into competition with Robert Stephenson, when a prize of £500 was offered for the best locomotive. He came off second-best, but it was a feather in his cap that his locomotive, the *Novelty*, glided smoothly over the track at the amazing speed of thirty miles an hour! His experiments with hot air occupied much of his time, and not without valuable results. His forte, however, was in the construction of steam-engines, of which he designed a large number, introducing many new principles, some of which were destined to survive.

Ericsson's first stroke of business in the United States made him famous. The *Princeton* war-ship (see page 69), built at the Philadelphia navy-yard under his direction, and fitted with his screw propeller, proved a great success, and gained him the favour and patronage of the government officials. Soon after the completion of the *Princeton*, he embarked in what he then accounted the greatest enterprise of his life—

THE CALORIC SHIP "ERICSSON."

With the financial assistance of several wealthy friends in New York, Ericsson proceeded to build a large sea-going vessel, to be propelled by means of hot air. It was a costly experiment, involving an outlay of $500,000, the engines alone costing $130,000. The cylinders were 168 inches in diameter, with six-feet stroke. The machinery was in motion within seven months of the laying of the vessel's keel. On her trial trip the *Ericsson* attained a speed of eight miles an hour, and subsequently as much as eleven miles an hour. The *Ericsson* was at once a success and a failure. She sustained the inventor's theory as to the power of heated air, *but* so excessive was the temperature of the air required to develop the power, the cylinders were warped out of shape and some of the fittings were burned to a crisp. The costly experiment was consequently abandoned. The caloric engine was replaced by an ordinary steam-engine, and thus transformed the *Ericsson* earned her living for many years.

THE "MONITOR."

This further product of Ericsson's fertile brain is in the form of an armour-protected, semi-submerged steam vessel for war purposes, and first came prominently into notice in connection with the memorable contest which took place in Hampton Roads on the 9th of March, 1862, between the *Merrimac* and *Monitor*. The former was an old wooden vessel refitted by the Confederate Government at Norfolk navy-yard, and covered with protective armour to the water-line. The *Monitor* was a flat iron boat resembling a scow, with nothing visible above water save the flush deck, from the centre of which rose a massive iron tower containing two guns of heavy calibre. The "cheesebox," as the *Monitor* was contemptuously styled, held her own against the *Merrimac*, which carried eleven guns. It was a drawn battle, but a victory for Ericsson, and resulted in many other steam vessels of this description being built for harbour and coast defence under his supervision.

John Ericsson died in New York on the 8th of March, 1889. *Vide* "Ericsson and His Inventions," in *Atlantic Monthly* for July, 1862, and "John Ericsson, the Engineer," in *Scribner's Magazine* for March, 1890.

II.

THE WHALEBACK

was invented and patented some years ago by Captain McDougall, of Duluth, a long-headed and level-headed Scotchman hailing from the famed island of Islay. The peculiarity of its construction consists in its elliptical form, combining strength of hull, cheapness of first cost and working, and large carrying capacity upon a light draught of water. Having no masts, the whaleback is entirely dependent on its steam-power, which in case of a breakdown or heavy weather renders the vessel helpless and unmanageable; but, on the other hand, it is contended that so long as she has sufficient water under her she is practically unsinkable. She has no deck to speak of, and consequently nothing to wash overboard save the waves, which play harmlessly over her arched roofing. Her hold is, so to speak, hermetically sealed. Though chiefly intended to carry freight, the capabilities of the whaleback as a passenger steamer have been satisfactorily tested. The *Christopher Columbus*, built on this principle, did duty as an excursion steamer at the Chicago World's Fair, and is now plying regularly as a passenger boat between Chicago and Milwaukee—the largest excursion steamer, so it is said, in the world, "having a carrying capacity of 5,000, which number of persons she has comfortably transported on a number of occasions." The steamer is 362 feet in length, has engines of 2,800 horse-power, and runs at the rate of twenty miles an hour. A considerable number of "whalebacks" are now engaged in the Upper Lakes grain and iron ore trade, all of them having been built by the Steel Barge Company at West Superior.

THE "JOHN S. COLBY" WHALEBACK.
From a photo presented by Mr. D. G. Thomson, of Montreal.

The above cut is a faithful representation of a type of steamer peculiar to the Upper Lakes, which, though somewhat odd-looking, is said to answer its purpose well as a grain-carrier.

The latest addition to the fleet is the biggest vessel of her class, and just now the largest grain-carrier on the lakes. This vessel, named after the inventor, *Alexander McDougall*, is 430 feet in length over all, 50 feet moulded breadth, and 27 feet in depth. Her double bottom is five feet deep, giving her a total water ballast capacity of 2,000 tons. Her displacement on a draught of 18 feet is about 10,000 tons, and she is able to carry the enormous cargo of 7,200 tons, equivalent to 240,000 bushels of wheat. She is built of steel, and has quadruple expansion engines. The only departure from the original whaleback in this instance is the substitution of the perpendicular stem for the "swinish snout" or "spoon bow," which has called forth so many uncomplimentary remarks, and which is much in evidence in our cut.

In 1891 the whaleback *Wetmore* was the first of this class of vessels to bring a cargo of grain from the Upper Lakes to Montreal and continue the voyage to Liverpool, where she arrived safely on July 21st. From Liverpool the *Wetmore* sailed to the Pacific coast *via* Cape Horn, and while carrying a cargo of coal from Puget Sound to San Francisco she was disabled in a violent storm, went ashore, and was wrecked.

III.

THE TURRET STEAMSHIP.

The hull of the turret ship closely resembles that of the whaleback, but instead of the "spoon bow" it has the straight stem, and is further distinguished by a "turret deck," so called, about one-third the width of the vessel and extending over its entire length, at a height of some five or six feet above the turn of the hull. This forms the working deck, and towering above it are the bridge, the cook's galley, the engineers' quarters, and other two-story erections, forming an unship-shapely *tout ensemble* of a most unprepossessing appearance; and yet, this is the type of steamship at one time seriously proposed by the contractors for the Canadian fast-line service! There are some thirty-five such vessels afloat in different parts of the world, all built at Sunderland, and most of them engaged in the coal trade, for which they are said to be well adapted.

The *Turret Age*, which plies between Sydney, C.B., and Montreal during the season of navigation, was built in 1893, and is owned by Messrs. Peterson, Tate & Co., of Newcastle on-Tyne. She is one of the largest of her class, being 311 feet in length, 38.2 feet in width, and 21.6 feet deep. She is propelled by a single screw, has a speed of eleven knots, and carries 3,700 tons of coal. Her capacious, unobstructed hold and continuous hatchway permit of loading and discharging cargo with marvellous rapidity, and she is said to be a fairly good sea-boat.

IV.

WATER-JET SYSTEM OF PROPULSION.

While Ericsson, Smith, Woodcroft and Lowe were busying themselves with experiments for perfecting the principle of the submerged screw as a means of propelling vessels through the water, another plan was being devised which, for a time, excited much interest, and was very nearly becoming a success. This was Ruthven's water-jet propeller. It differed from Ericsson's in the singular fact that the actual propeller was placed inside of the ship instead of on the outside. This propeller, in the shape of a fan-wheel with curved blades, was made to revolve horizontally and rapidly in a tank of water placed in the hold of the vessel, fed from the sea through openings in the hull. The power of the steam-engine was applied to expelling the water from this tank through curved pipes with nozzles, on either side of the ship. In proportion to the velocity with which the water was forced through these pipes into the sea below the water-line, an impetus in the opposite direction was given to the vessel. The nozzles were so constructed that they could be turned easily towards the bow or stern, as occasion required, for forward or backward motion. The first experiment with this appliance was made by Messrs. Ruthven, of Edinburgh, on the Frith of Forth, with an iron boat 40 feet in length, in 1843, when a speed of seven miles an hour was attained. The *Enterprise*, 90 feet long and 100 tons burthen, was built on this principle, and made her trial trip, January 16th, 1854, when she developed a speed of 9.35 miles an hour. This vessel was intended for the deep-sea fishing, and the jet-propeller was suggested in this case as being less liable to become entangled with the nets than the screw or paddle. The water-jet system was also tried on a Rhine passenger steamboat with some measure of success; but while the theory was upheld, it seems to have failed in practice, because the results in speed and in other respects were not proportioned to the working power and the consumption of fuel. See *En. Britannica*, 8th ed., vol. xx., p. 661.

V.

THE CIGAR STEAMBOAT.

Experiments with this style of river craft have been frequent on both sides of the Atlantic without, however, being followed by substantial success. So long ago as 1835, the *Rapid*, consisting of two hollow cylinders, pointed at either end in cigar fashion, placed ten feet apart, with a large wheel between them in the centre, appeared on the Upper St. Lawrence, fitted with the steam-engine of the superannuated *Jack Downing*. Her first trip down the river was also her last, for, after many fruitless attempts to return, she was wrecked, and for a time abandoned. Eventually, she was towed, by way of the Ottawa and Rideau canals, to Ogdensburg, where she was refitted and plied for some time as a ferry boat. A very pretty specimen of a cigar-boat built of iron, with an elegant superstructure, the writer remembers having seen on the Clyde more than half a century ago, but as to its career and ultimate fate deponent sayeth not. A twin-boat steamer, reminding us of Patrick Miller's first attempt at steam-boating, propelled, however, by side-wheels, may be seen any day during the season of navigation dragging its slow length along on the ferry from Laprairie to the opposite shore of the St. Lawrence, near Montreal.

VI.

THE ROLLER STEAMBOAT.

The reader is requested to put on his thinking cap before endeavouring to comprehend the brief reference now to be made to Mr. Knapp's "Roller." On the 8th of September, 1897, there was launched from the yard of the well-known Polson's Iron Works Company in Toronto, an enlarged model of the strangest craft ever seen—a huge innovation upon all preconceived ideas of marine architecture. The exterior of the boat in question, if it can be called a boat, has all the appearance of a round boiler 110 feet long and 25 feet in diameter. The outer cylinder is built of one-quarter inch steel plates stoutly ribbed and riveted, and armed with a number of fins, or small paddles, the ends being funnel-shaped, with openings in the centre. This is made to revolve by means of two engines of 60 horse-power each, placed one at either end of the vessel. An inner cylinder similarly constructed, corresponding to the hold of a ship, remains stationary while the other is supposed to be rolling over the surface of the water, regardless of wind and waves, at railway speed. The modest calculation of the inventor is that a steam vessel so constructed of 700 feet in length and 150 feet in diameter, *ought* to cover the distance between New York and Liverpool in forty-eight hours! This model was built at a cost of $10,000. The results of the trial trip on Toronto Bay have not been made public.

VII.

THE "TURBINIA."

In June, 1897, there appeared on the Solent, at the time of the great Jubilee Naval Review, a steam vessel furnished with a novel method of propulsion, by which a speed far in excess of any previous record was attained. In the opinion of competent experts this new application of steam-power is likely to bring about in the near future a revolution in steam navigation. The following account of this phenomenal craft appeared in the Montreal *Star:*

"LONDON, July 5th, 1897.

"The record-breaking 100-foot torpedo boat *Turbinia* has intensely interested the public here generally, and experts in marine engineering in particular. It is admitted that if the principle of the steam turbine invented by Charles Parsons and fitted in the *Turbinia* can be extended to large ships, it will mark the greatest revolution in mechanics since the invention of the steam-engine itself.

"Mr. Wolff, M.P. for Belfast, head of the famous firm of Harland & Wolff, of Belfast, and himself the designer of the White Star Liners, says :

"'I saw the *Turbinia* at Spithead going nearly eight miles an hour faster than any vessel had ever gone before, and even then she was not being pushed to her full speed. She passed quite close to the *Teutonic*, on which I was. She dashed along with marvellous speed and smoothness.

"'I must say, however, that I felt more secure on the *Teutonic* than I should have felt on the *Turbinia*, for you know they have not yet surmounted the difficulty of reversing the engine. She can go ahead forty miles an hour but can only reverse at less than four.

"'If Parsons can make a similar turbine engine practicable for big craft with proper reversing power, he will open a new era in the history of steam motors. But, although he has carried the economizing of steam to a great pitch for a turbine engine, still from my observation the waste of both steam and fuel under his system, if applied on a large scale, would be almost fatal. That there is a big future before his turbine engine for launches and other small craft I do not doubt, provided that he can get over the reversing difficulty.'"

The *Scientific American,* in its issue of June 26th, 1897, says: " Nothing more startling has ever occurred than the wonderful runs which have recently been made by a little craft called the *Turbinia,* in which the motive power is supplied by a steam turbine of the Parsons type."

Quoting from a paper read at a meeting of the Institution of Civil Engineers in London, by the Hon. Charles A. Parsons, the inventor of this new system, the advantages of the turbine system are thus summarized:

"(1) Greatly increased speed, owing to diminution of weight and smaller steam consumption; (2) increased carrying power of vessel; (3) increased economy in coal consumption; (4) increased facilities for navigating shallow waters; (5) increased stability of vessel; (6) reduced weight of machinery; (7) reduced cost of attendance on machinery; (8) reduced size and weight of screw propellers and shafting; (9) absence of vibration; (10) lowered centre of gravity of machinery, and reduced risk in time of war.

"The *Turbinia* is 100 ft. in length, 9 ft. beam, 3 ft. draught amidships, and $44\frac{1}{2}$ tons displacement. She has three screw shafts, each directly driven by a compound steam turbine of the parallel flow type. The three turbines are in series, and the steam is expanded—at full power—from a pressure of 170 pound absolute, at which it reaches the motor, to a pressure of one pound absolute, at which it is condensed. The shafts are slightly inclined, and each carries three screws, making nine in all. The screws have a diameter of 18 in., and when running at full speed they make 2,200 revolutions per minute. Steam is supplied from a water tube boiler, and the draught is forced by a fan, mounted on the prolongation of the low pressure motor shaft, the advantage of this arrangement being that the draught is increased as the demand for steam increases, and also that the power to drive the fan is obtained directly from the main engines.

"Up to the present the maximum mean speed attained has been $32\frac{3}{4}$ knots, as the mean of two consecutive runs on the measured mile. These runs were made after about four hours' steaming at other speeds, and the boat on the day of the trials had been fifteen days in the water. It is anticipated that on subsequent trials, after some alterations to the steam pipe, still higher mean speeds will be obtained.

"It is believed that when boats of 200 feet in length and upward are fitted with compound turbine motors, speeds of 35 to 40 knots may be easily obtained in vessels of the destroyer class, and it is also believed that the turbine will—in a lesser degree—enable higher speeds to be realized in all classes of passenger vessels."

Referring to the difficulty of reversing the engines of the *Turbinia*, the *Scientific American* adds, that "by using a system of 'butterfly' reversing steam valves, a motor has been constructed in which the steam may be made to flow through the blades of the turbine in either direction, the whole horse-power of the engines being thus available for going astern." Detailed drawings and descriptions of the *Turbinia* and the new motor may be found in the supplements of the *Scientific American* (New York) for June 26th, 1897, and March 12th, 1898.

INDEX.

Letter "S" indicates Inland Steamer, "SS" Ocean Steamer.

Aberdeen Steamship Line, 156.
Acadia, SS., 73.
Accommodation, S., 50, 312.
Adriatic, SS., Collins, 105.
Adriatic, SS., White Star, 118.
African Steamship Company, 154.
Aird, Captain, 215.
Aitken & Company, steamship builders, 286.
Alaska, SS., 116.
Albany to Montreal, 260.
Alberta, S., 284.
Algoma, S., 255, 284.
Allan, Alexander, 196, 209.
Allan, Andrew, 196, 296.
Allan, Bryce, 196, 209.
Allan, James, 196, 209.
Allan, Sir Hugh, 196, 208.
Allan Steamship Line, 196.
Alps, SS., 99.
Amazon, steel barge, 302.
America, SS., 114.
Amerika, SS., 141.
American Steamship Line, Lake Ontario, 327.
Anchor Steamship Line, 113, 151.
Ancient, Rev. W. J., 122.
Anderson, Captain, 86.
Angloman, SS., 225.
Anglo-Saxon, SS., wrecked, 199.
Appomattox, S., 272.
Archer, Captain, 202, 213.
Archimedes, S., 68.
Arctic, SS., 104, 106.

Arizona, SS., 116.
Armed cruisers, 172.
Armed mail packets, 73.
Arrow Steamship Line, 129.
Athabaska, S., 284.
Athenian, SS., 164.
Atlantic, SS., Collins, 104-106.
Atlantic, SS., White Star, 121.
Atlantic Transport Steamship Line, 129.
Augusta Victoria, SS., 132.
Australasian, SS., 88.
Australia, SS., P. & O., 147.
Australia and Vancouver Steamship Line, 164.
Austria, SS., burned, 134.
Aylmer, Lord, 54.

Bain, Captain Robert, 36.
Ballantine, Captain, 200.
Baltic, SS., Collins, 104-106.
Baltic, SS., White Star, 118.
Bannockburn, S., 286, 293.
Barbadian, SS., 157.
Barber & Company Steamship Line, 129.
Barber, Keith A., 343.
Barclay & Curle, builders, 205.
Battleships, 171.
Bay of Fundy, 188.
Beauharnois Canal, 265.
Beaver Steamship Line, The, 229.
Beaver, The old steamer, 334.
Belgravia, SS., 113.

INDEX.

Bell, Henry, 36.
Bibby Steamship Line, 151.
Black Ball Steamship Line, 27.
Black Diamond Steamship Line, 235.
Blue Flag Steamship Line, 129.
Bohemian, SS., 199; wrecked, 202.
Boothby, Captain, 186.
Boulton & Watt, engineers, 334.
Brandon to Britain, 295.
Bristol City Steamship Line, 129.
Britannia, SS., 72, 74.
Britannic, SS., 118.
British and African Steamship Company, 155.
British and Colonial Steam Navigation Company, 156.
British Columbia, 334.
British India Steam Navigation Company, 148.
British and North American Royal Mail Steam-Packet Company, 73.
British navy, 166, 175.
British Queen, SS., 97.
Brooks, Captain, 102.
Brown, Captain, 216.
Bruce Mines, S., 254.
Brunel, Isambard, 66.
Brush, George, 307, 310.
Buenos Ayrean, SS., 206.
Bulwer, Sir Edward, 159.
Burial of dead at sea, 183.
Burlington, S., 44.
Burns, Rev. Dr., 94.
Burns, Sir George, 71, 93.

Calcutta and Burmah Steam Navigation Company, 148.
Caledonia, SS., Cunard, 73.
Caledonia, SS., P. & O., 146.
Calvin Company, 287.
Cameron, Captain, 123.
Campana, S., 235.
Campania, SS., 78, 174.
Campbell, Captain Howard, 234.
Canada, SS., Cunard Line, 75.
Canada, SS., Dominion Line, 226.
Canada Shipping Company, 229.

Canadian, SS., 198-200.
Canadian canals, 258.
Canadian commerce on lakes, 283.
Canadian Pacific Railway, 158.
Canadian Pacific steamers, 160, 164, 284.
Canadian Steam Navigation Company, 316.
Canal tariffs, 303.
Cape of Good Hope, SS., 149.
Car of Commerce, S., 310.
Carthaginian, SS., 206.
Castle Steamship Line, The, 155.
Celtic, SS., 118.
Charity, SS., 195.
Charlotte Dundas, S., 33.
Chesapeake and Ohio Steamship Line, 129.
Chicora, S., 255.
Chieftain, S., 326.
Chimborazo, SS., 148.
China, SS., 75.
Chippewa, S., 254.
Cimbria, SS., sunk, 134.
Circassia, SS., 186.
Circassian, SS., 205.
City of Berlin, SS., 108.
City of Boston, SS., 107.
City of Brussels, SS., 107.
City of Chicago, SS., 107.
City of Glasgow, SS., 107.
City of Manchester, SS., 107.
City of Montreal, SS., 107.
City of New York, SS., 108.
City of Paris, SS., 108.
City of Philadelphia, SS., 107.
City of Rome, SS., 113, 128.
City of Washington, SS., 107.
City Steamship Line to India, 152.
Clan Steamship Line, The, 150.
Cleopatra, SS., 195.
Clermont, S., 41.
Cleveland, Ohio, 278, 281.
Clipper ships, 26.
Clyde River steamers, 38.
Codfish industry, 355.
Collingwood and Owen Sound, 255.
Collins, E. K., 106.
Collins Steamship Line, 99, 103.

INDEX. 375

Collision at sea, 126.
Columba, S., 38.
Comet, S., Bell's, 34, 312.
Commerce of Great Lakes, 268.
Compagnie Generale Transatlantique, 138.
Compound engines, 100, 345.
Connal & Co., builders, 222.
Continental Steamship Lines, 130.
Cook, Captain, 86, 88.
Corona, S., 330.
Cost of running steamships, 84.
Cramp & Sons, builders, 110.
Crathie, SS., collision, 136.
Crescent, H.M.S., 189.
Crimean War, 198, 214.
Cruisers, Armed, 172.
Cumberland, S., 255.
Cunard fleet, 85.
Cunard Steamship Line, 73.
Cunard, Sir Edward, 93.
Cunard, Sir Samuel, 71, 91.
Cunard track chart, 96, 176.
Currie, Captain, 207.
Cushing, Manager, 318.
Cuzco, SS., 148.

Dakota, SS., 115.
Dalziel, Captain, 203.
Danmark, SS., foundered, 141.
Danube, SS., 157.
Dawn of steam navigation, 28.
Deeper waterways, 299, 302.
Dennys, ship-builders, 154, 198, 204.
Detroit River tonnage, 276.
Devonia, SS., 113.
Diamond Jubilee Review, 170.
Dick, Captain, 324.
Dickens, Charles, 18.
Distances, Marine, 177.
Dolphin, S., 325, 326.
Dominion Steamship Line, 221.
Dominion Steamers, 353.
Donaldson Steamship Line, 234.
Douglas, Captain, 75.
Douglas, Governor of British Columbia, 336.
Dramatic Line, The, 103.
Draught, Induced, 20.

Drummond Castle, SS., lost, 155.
Dry-docks, 342.
Duke of Marlborough, H.M.S., 168.
Duke of Wellington, H.M.S., 97, 168.
Durham boats, 260.
Durham City, SS., 190.
Dutton, Captain, 217.

Early Atlantic steamers, 50.
Eastern trade, The, 153.
East India Company, 142.
Elbe, SS., sunk, 136.
Elder, Dempster Steamship Line, 156, 235.
Elder, John, & Co., 100, 116, 132.
Eldridge, Captain, 106.
Elevator, The grain, 290.
Emerald, S., 254.
Emigrant ships, 20, 210.
Empress Steamship Line, 160.
Empire, S., 255.
Empire City, S., 271.
Enterprise, SS., 53.
Ericsson, John, inventor, 67.
Erie Canal, 280.
Erin, SS., lost, 115.
Etolia, SS., in the ice, 185.
Etruria, SS., 77, 119, 189.
Europa, SS., 75.
European, SS., 157.
Eutopia, SS., sunk, 114.
Evans, Captain, 185.
Exports from Montreal, 267.

Fares to India and the East, 147, 153.
Fairfield Ship-yard, 78, 100, 346.
Farlinger, Captain, 327.
Fast Line of Steamships, 236, 242.
Fast service to Japan, 156.
Favourite, sailing-ship, 196.
Fawcett, William, SS., 146.
Ferry-boats, American, 48.
First compound engine, 345.
First live stock shipment, 236.
First lake propeller, 252.
First steamer in Canada, 50, 312.

376 INDEX.

First steamer on Lake Ontario, 247.
First steamer on Lake Erie, 251.
First ocean steamship, 54.
First steam fog-horn, 347.
First steel steamship, 206.
First wheat shipment from Manitoba, 295.
Fleming, Sir Sandford, 159, 239, 242.
Floating elevators, 295.
Flying Squadron, The, 170.
Fox, Sir Douglas, 144.
Francis B. Ogden, S., 68.
Francis Smith, S., 255.
Frederick the Great, SS., 144.
Freight, inland rates, 303.
French Steamship Line, 138.
Friesland, SS., 113.
Frontenac, S., 247.
Fulda, SS., 86, 136.
Fulton, Robert, 41.
Furnessia, SS., 113.
Furness Steamship Line, 235.
Furst Bismarck, SS., 131.

Gallia, SS., 234.
Garonne, SS., 148.
Gaskin, Captain, 263.
General Smyth, S., 343.
Genova, SS., 195.
German East African Steamship Line, 156.
Germanic, SS., 118, 127.
Gildersleeve, S., 320.
Gildersleeve, Manager, 316.
Glenmorag, ship, wrecked, 207.
Golconda, SS., 149.
Gore, S., 254.
Gothic, SS., 151.
Graham, Captain John, 210.
Grain-sucker, 291.
Grain elevator, 290.
Grand Trunk Railway opened, 328.
Grange, Captain, 209.
"Graphic," The London, 171.
Graving-docks, 342.
Great Britain, SS., 61.

Great Eastern, SS., 62.
Great Lakes, The, 244.
Great Northern Transit Company, 288.
Great Republic, SS., 26.
Great Western, SS., 60.
Great Western Railroad Line, 327.
Grenville Canal, 318.
Griffin, schooner, 246.
Guion Steamship Line, 115.
Gulf ports, Map of, 241.

Hagart & Crangle Line, 287.
Haines, Captain, 89.
Haliburton, Judge, 93, 159.
Halifax harbour, 340.
Hall Steamship Line, 152.
Hamburg & American Steamship Packet Company, 130.
Hamilton, Captain Clarke, 327.
Hamilton, Hon. John, 323, 331.
Hamilton, S., 327.
Hamilton Steam Navigation Company, 330.
Handyside & Henderson, 113.
Hansa St. Lawrence Steamship Line, 235.
Harland & Wolff, 117, 123, 140, 151, 228.
Harrison, Captain, 86.
Havel, SS., 137.
Head Steamship Line, 235.
Henderson Steamship Line, 152.
Hennepin, Father, 246.
Hercules, S., 252, 309.
Hibernia, SS., 87.
Hibernian, SS., 204.
Highlander, S., 324.
Hill Steamship Line, 129.
Himalaya, SS., 147.
Hindostan, SS., 146.
Hooker & Jones, forwarders, 318.
Hornet, torpedo destroyer, 169.
Horse-boat, The, 29.
Howard, Captain Thomas, 320, 327, 328.
Howe, Hon. Joseph, 159.
Howland, O. A., 301.

INDEX. 377

Hudson's Bay Company, 332, 333.
Hungarian, SS., lost, 199, 200.

Icebergs, 183.
Idaho, SS., lost, 225.
Imrie, William, 117.
Independence, propeller, 257.
India, SS., 149.
India and the East, 142.
Indian, SS., 142, 198, 200.
Indiana, SS., U.S., 342.
Inman Steamship Line, 107.
International Steamship Line, 107, 109.
Inverclyde, Lord, 94, 99.
Ireland, propeller, 263.
Iron steamers, 61, 314.
Iron ore transportation, 279.
Iroquois, S., 326.
Ismay, Thomas, H., 116, 122.

James Swift, S., 331.
James Watt, S., 271.
John Jacob Astor, sail vessel, 256.
John Kenzie, brig, 254.
John Munn, S., 313.
Johnston Steamship Line, 235.
Jones, Captain J., 66, 202.
Jones, Captain Thomas, 209.
Jones, J. & J., forwarders, 318.
Jubilee Review, 170.
Judkins, Captain, 86.
Julia Palmer, propeller, 257.
Jura, SS., stranded, 202.

Kaiser Wilhelm der Grosse, SS., 136.
Kaiser Wilhelm II, 136.
Keefer, Thomas, C. E., 283, 301.
Kent, S., 254.
Kingsford, Historian, 263, 283.
Kingston, Ontario, 331.
Kingston, S., 327.
Klondike, Steam to, 164.

Labrador, SS., 223.
Lachine Canal, 259.
Lady Colborne, S., 314.

Lady Eglinton, S., 195.
Lady Elgin, S., 314.
Lady Sherbrooke, S., 310, 312.
Lady Washington, schooner, 247.
Lahn, SS., 136.
Lake Ontario, SS., 231.
Lake St. Peter, 266.
Lake Superior, SS., 231.
Lakes, Navigation Companies, 270.
Lakes, The Great, 244.
La France, ship, 28.
La Salle, explorer, 246.
La Bourgogne, SS., lost, 138.
La Touraine, SS., 138.
Lamport & Holt Steamship Line, 129, 157.
Life-boats at sea, 125.
Lindall, Captain, 222.
Live stock exportation, 236.
Liverpool landing-stage, 81.
Liverpool packet-ships, 27.
Liverpool, SS., 58.
Lochearn, SS., collision, 140.
Locomotives, 294.
Lord Steamship Line, 129.
Lord Sydenham, S., 314.
Lott, Captain, 86, 88.
Lowe, James, inventor, 68.
Lucania, SS., 78.
Lusitania, SS., 148.

Magnet, S., 327.
Majestic, SS., 119.
Malsham, S., 310.
Manchester Ship Canal, 235.
Manhanset Steamship Line, 129.
Manitoba, S., 286.
Manitou, S., 270.
Map of the Gulf of St. Lawrence, 241.
Marjery, S., 40.
Marine distances, 175.
Mariposa, SS., wrecked, 225.
Marshall, Captain, 320.
Matiana, SS., 149.
Maudsley, Field & Company, engineers, 118.
Memphis, SS., lost, 235.

Merchant Lines, Hamilton, 287.
Merritt, Hon. William, 262.
Messageries Maritimes Steamship Company, 153.
Miller, Patrick, 31.
Milloy, Alexander, 316.
Miowera, SS., 164.
Missouri, SS., 141.
Moldavia, SS., 186.
Molson, Hon. John, 307.
Monarch, S., 287.
Montana, SS., 115.
Montreal Ocean Steamship Company, 198.
Montreal, Port of, 266.
Montreal steamer burned, 315.
Montreal Transportation Company, 286.
Moodie, Captain, 86.
Moravian, SS., wrecked, 202.
Morris, Hon. Alex., 159.
MountStephen, Lord, 164.
Munro, Thomas, C. E., 301.
Murrell, Captain, 141.
Mutiny at sea, 24.
Macaulay, Captain, 227.
Macdougall, Captain John, 57.
Maclean, Captain N., 217.
Macleod, Dr. Norman, 179.
Macpherson, Crane & Co., 318.
McIver, David, 71, 95.
McKean, McLarty & Co., 195.
McKenzie, Captain, 248.
McKinstry, Captain, 127.
McLennan, Hugh, 296.
McMaster, Captain, 209.

Napier, David, 35.
Napier, Robert, 96, 71, 148, 168, 205.
Napoleon, S., 314.
Naronic, SS., lost at sea, 122.
Natal Steamship Line, 156.
National Steamship Line, 114.
Navy, The Royal, 166, 175..
Nestorian, SS., 205.
Netherlands Steamship Line, 140.
New England, SS., 229.
Newfoundland, 354.

New York, SS., 108, 111.
Niagara Ship Canal, 302.
Niagara, SS., 74.
Niagara Steam Navigation Company, 329.
Nile, SS., 157.
Norman, SS., 155.
Normannia, SS., 131.
North Atlantic Steamship Company, P. E. I., 349.
North American, SS., 199.
North Briton, SS., lost, 202.
North American Transport Company, 129.
Northern Light, S., 351.
Northern Steamship Company, 272.
North German Lloyd Steamship Company, 134.
North Shore Navigation Company, 288.
North-West Fur Company, 256, 277.
North-West Navigation Company, 333.
North-West, S., 274.
North-West Transportation Company, 287.
Norwegian, SS., wrecked, 202, 204.
Nova Scotia, 340.
Nova Scotian, SS., 199.

Ogilvie, W. W., 297.
Oldfield, S., 321.
Old Man of the Sea, 102.
Ontario Lake Navigation, 328.
Ontario, S., 248, 326.
Ontario, SS., 222.
Ophir, SS., 148.
Oregon, SS., Cunard, sunk, 86, 99.
Oregon, SS., Dominion, 222.
Orient Steam Navigation Company, 147.
Orizaba, SS., 157.
Ottawa, SS., 195, 225.
Ottawa and Georgian Bay Canal, 304.
Ottawa and Rideau Forwarding Company, 310, 318.

INDEX.

Ottawa River steamers, 321.
Ottawa River Navigation Company, 318.
Overland route, The, 143.
Owego, S., 270.

Pacific, SS., 104, 106.
Pacific Steamship Navigation Company, 157.
Packet-ships, 27.
Papin, Denis, 20.
Paris, SS., 108, 125, 189.
Parisian, SS., 205.
Parsell, Captain, 123.
Passport, S., 327.
Patterson of Bristol, 60.
Paynter, George, 102.
Penelope, H.M.S., 168.
Peninsular and Oriental Steamship Company, 145.
Pennsylvania, SS., 101, 134.
Persia, SS., 75, 97.
Peruvian, SS., 205.
Peterson, Tate & Co., 237.
Phœnician, SS., 207.
Pioneer, S., 252.
Ploughboy, S., 254.
Polynesian, SS., 205.
Pomeranian in a storm, 203.
Pomone, French war-ship, 69.
Postal compensation, 132.
President, SS., lost at sea, 61.
Prince Edward Island, 347.
Prince of Wales, war-ship, 168.
Princeton, war-ship, 69.
Priscilla, S., 44.
Provisions, Ships', 83.
Puffers, 319.
Pumper, S., 264.

Quebec Province, 307.
Quebec and Halifax Steamship Company, 66.
Quebec, S., 311.
Quebec Steamship Company, 235.
Queen Charlotte, S., 249.
Queen City, S., 293.
Quetta, SS., wrecked, 149.

Racing at sea, 125.
Randolph, Elder & Co., 100.
Rates of passage, 124.
Rathbun Company, 330.
Rattler, H.M.S., 69.
Recovery, brigantine, 256.
Red Star Steamship Line, 112.
Renown, H.M.S., 172.
Republic, SS., White Star, 118.
Richardson, Captain, 217.
Richard Smith, S., 347.
Richards, Mills & Co., 224.
Richelieu Steamboat Company, 314.
Rideau Canal, 264.
Ritchie, Captain, 216.
Robert Garrett, S., 48.
Rob Roy, S., 40.
Rockefeller Fleet, 271.
Rosemount, S., 286.
Royal Mail West Indies Steam-Packet Company, 156.
Royal William, S.S., 54, 340, 347.
Rubattino Steamship Line, 153.
Russell, Scott, 63.
Russia, SS., 75.

Sail *versus* Steam, 247.
Salier, SS., lost at sea, 136.
Sampson, propeller, 252.
Sam Ward, S., 257.
Sarah Sands, SS., 195.
Sardinian, SS., 205, 217.
Sarmatian, SS., 198.
Sarnia, SS., 222.
Sault Ste. Marie Canal, 276.
Savannah, SS., 51.
Scotia, SS., 75, 97.
Scotsman, SS., 225.
Scott & Company, 138.
Schiller, SS., wrecked, 134.
Screw propeller, The, 67.
Sealing steamers, 355.
Servia, SS., 76.
Shaw, Savill and Albion Steamship Company, 151.
Shenango, ferry steamer, 49.
Shepherd, Captain H. W., 322.
Shepherd, Captain R. W., 321.

380 INDEX.

Ship-building, 279.
Ship canals, 303.
Siberian, SS., 206.
Simpson, Sir George, 258.
Simcoe, General, 258.
Sirius, SS., 59.
Sir Robert Peel, S., 324.
Smith, T. P., inventor, 67.
Smith, Captain W. H., 194, 214.
Smith, Donald A., 159.
Smythe, Major C., 158.
Sophia, S., 249.
Sovereign, S., 317.
Spaarndam, SS., 141.
Spitfire, H.M.S., 354.
Spithead reviews, 173.
Spree, SS., 136.
Stanley, S., P. E. I., 351.
State Steamship Line, 129.
Steam Navigation in British Columbia, 334.
Steam Navigation in New Brunswick, 343.
Steam Navigation on the Ottawa, 317.
Steam Navigation in Newfoundland, 354.
Steam Navigation in Nova Scotia, 340.
Steam Navigation in Prince Edward Island, 347.
Steam Navigation in Quebec, 307.
Steam Navigation in Manitoba, 332.
Steam Navigation in Ontario, 323.
Stearns, Captain, 324.
Steel barges, 282.
Steel steamships, First, 206.
Stephen, George, 159, 164.
Stewart, Macleod, 304.
Stone, Captain, 86.
Strachan, Bishop, 21.
St. George, SS., wrecked, 202.
St. John harbour, N. B., 345.
St. Lawrence canals, 258, 264.
St. Lawrence route, 192.
St. Mary's Falls Canal, 276, 278.
St. Louis, SS., 110.
St. Paul, SS., 110.

Strathcona, Lord, 159, 164.
Subsidies to steamship companies, 104, 111, 161.
Subventions, 120.
Suez Canal, 144, 149.
Summary of Steam Navigation, 356.
Sunday at sea, 178.
Sutherland, Captain, 327.
Swearing, Profane, 220.
Swiftsure, S., 310.
Symington, William, 31.

Tartar, SS., 164.
Taylor, T. F., 284.
Taylor, Dr. W. M., 179.
Tate Brothers, builders, 314.
Thingvalla Steamship Line, 141.
Thomas MacKay, S., 320.
Thomson, J. A., steamboat inspector, 334.
Thomson Steamship Line, 235.
Thomson, J. and G., steamship builders, 108, 113, 123.
Teutonic, SS., 119, 174.
Tidal waves, 188.
Tod & McGregor, engineers, 107.
Tonnage on the Great Lakes, 276.
Toronto and Steam Navigation, 329.
Torpedo boats, 169.
Torrance, John, 228, 308.
Torrance, Messrs. David, & Co., 221, 307.
Transportation companies, 284.
Transportation business, 289.
Trave, SS., 136.
Trent, SS., 88.
Trevethick, Engineer, 67.
Tripoli, SS., lost, 86.
Twohey, Captain, 324.

Ulster Steamship Company, 235.
Umbria, SS., 77, 119.
Unicorn, SS., 75.
Union Steamship Company, Africa, 154.
Union Steamship Company, New Zealand, 151.

United Empire, S., 287.
United Empire Loyalists, 258, 296.
United Kingdom, SS., 40.
United States Shipping Company, 129.
Up-to-date steamships, 18.
Utica, barge, 270.

Vancouver Island, 336.
Vancouver, SS., 222.
Vandalia, propeller, 252.
Vesta, SS., 106.
Vicksburg, SS., lost, 224.
Victoria, B.C., founded, 336.
Victoria Steamboat Association, 38.
Ville de Havre, SS., lost, 140.
Ville de Ciotat, SS., 153.
Voyageurs, Early, 258.

Waghorn, Lieut., 143.
Waldensian, SS., 207.
Walk-in-the-Water, S., 251.

Ward & Co., 310, 311.
Waring, Captain W L., 345.
Warrimoo, SS., 164.
Warrior, H.M.S., 168.
Washington, schooner, 246.
Waterways of Canada, 244.
Watt, James, engineer, 67.
Welland Canal, 262.
West Indies and Pacific Steamship Lines, 156.
Whale captured, 312.
White Star Steamship Line, 116.
William Fawcett, SS., 146.
William IV., S., 324.
Williams, Captain, 122.
Wilson Connoly Company, 313.
Wilson Steamship Line, 128.
Winter Ferry, P. E. I., 349.
Woodcroft, Engineer, 67.
Woodruff, Captain, 74.
World's Steamers, 357.
Wylie, Captain, 212.

Young, Captain, 128.

www.ingramcontent.com/pod-product-compliance
Lightning Source LLC
Chambersburg PA
CBHW051744300426
44115CB00007B/688